ASTRONOMY AND GEOLOGY

COMPARED.

BY

LORD ORMATHWAITE.

NEW YORK:
D. APPLETON & COMPANY,
549 & 551 BROADWAY.
1872.

AMERICAN PUBLISHERS' NOTE.

The author of the following work is an English Statesman who was born in 1798, and is consequently now seventy-four years of age. He served a long time in the House of Com mons under the title of Sir John Walsh, having succeeded his father as Baronet in 1825. He was M. P. for Sudbury in 1830–'34 and again in 1838–'40, and was M. P. for Radnorshire from 1840 to 1865, when he retired from public life. In 1868 he was raised to the peerage with the title of Lord Ormathwaite. In politics he is a Liberal Conservative. In a note prefixed to the English edition of his book he says: "I must preface these pages by a few words of explanation. The decay of my eyesight has within the last year incapacitated me from reading or writing. Cut off by this affliction from my usual mental employments, I have found relief and occupation to my thoughts in dictating these essays to a short-hand writer. This mode of writing will excuse any inaccuracies or mistakes which I have committed."

CONTENTS.

ASTRONOMY AND GEOLOGY

COMPARED.

There is a certain affinity between these two sciences: they have this feature in common—that each in its respective province extends the intellectual vision of Man far beyond the limits of his present mortal existence. They enable him to comprehend a vast portion of those two great infinities of time and of space by which he is surrounded. How wonderful that a being so ephemeral, whose life is compressed within the bounds of some seventy or eighty years at the utmost, and is cribbed and tethered to a narrow spot on this earth's surface, should thus be enabled to traverse in thought such a boundless area! Our first sentiment is perhaps one of melancholy and discouragement. When we contrast our own nothingness with the immensity of the Universe, which these sciences disclose to us, how infinitesimally small we appear in our own eyes; what mere atoms we are, no more permanent than the flies which buzz through a summer and then die; no larger than the motes which dance in the sunbeam! But our second thoughts are more inspiring and consolatory. Is not the fact that we are thus enabled to measure and to comprehend so large a part of this mighty scheme, a proof that we are connected with it by ties more lasting than are at present visible to us? The acts of the Almighty are never purposeless, and the circumstance that we are thus permitted a prospect of what is beyond us here, is one evidence, among the many others which are afforded to us, of the truth of those hopes which associate us with a larger future.

Not only do these two sciences afford us these prospects beyond the boundaries of this present world, but I think it will be found that they are the only ones which do so. The other sciences, which add so largely to our knowledge of the laws of nature here, and place in our hands such vastly augmented powers over the material world, deal almost exclusively with sublunary matters. Chemistry, Electricity, Anatomy, Botany, and many others, have their relations here. Other sciences borrow, indeed, from Astronomy or Geology, but they are merely dependent on them; they are secondary and subsidiary in their relations to them.

For example, the two sciences of Navigation and of Geography are largely indebted to Astronomy, which points out to them the modes, by the means of celestial observations, of verifying all the positions on the earth's surface; but Geography and Navigation only borrow from Astronomy the means by which they make these calculations, they add nothing to them, they only derive aid from them, they originate nothing.

Although, as I have remarked, there is a close connection, and, if I may use the expression, a sympathy, between these two great sciences, yet, on the other hand, there are several important differences between them which it may be interesting to observe. Let us endeavor to enumerate some of the most striking and remarkable.

First. The dates of their birth differ as widely as possible. Astronomy is about the oldest; Geology is nearly the youngest and last in the whole circle of the sciences. We must carry our knowledge of the commencement of Astronomy to the very beginning of the historic period; we must mount up to the days of the Assyrian monarchy, and the periods of the glories of Nineveh and Babylon; we must picture to ourselves the astronomers of that period watching the celestial firmament in the clear, dark blue skies, and through the transparent atmosphere, of Southern Asia. Nor let us too hastily depreciate, or too scornfully despise, the labors of these early pioneers on the road of Astronomical science. True, they supposed the earth the great centre of the universe, and conceived that sun, moon, and stars made a diurnal revolution round the earth, for the sole purpose of warming and enlightening us. True, their conceptions of the mechanism of the heavens, or of the relative magnitude of the great bodies which people the void, were utterly erroneous. Still we must remember that science advances step by step; they were very accurate observers of all they saw; they noted with great truth and precision all the apparent phenomena of the heavens; they mapped out very distinctly all the constellations which spangle the vault of heaven; they noticed very correctly all the signs of the zodiac; and to this day, although we know how to interpret all the motions for which they wanted the key, yet still we are contented to adopt their fanciful map of the constellations, and to use their nomenclature. Sir George Cornewall Lewis, in his interesting work on the "Astronomy of the Ancients," has done justice to their labors. Even, however, if we determine to ignore all these early periods, if Pythagoras and Ptolemy groped and blundered amidst confusion and error, and if we date the birth of true Astronomy from the days of Copernicus and Galileo, Kepler and Newton, it still has a priority in date of wellnigh four hundred years over its sister Geology, which cannot be said to have existed before the very end of the last and the commencement of the present century.

Another difference between the two sciences arises from the periods to which they refer. Geology is essentially a retrospective science; it never looks forward, it invites us to cast our regards behind us, and to endeavor through the obscurity of the past to trace out the former conditions of the earth's surface; its pages are all written in the præterpluperfect tense; just as some wizard or necromancer in the legends of the middle ages evoked by his potent spells the spirits of departed mailed knights, beauteous dames, or crowned monarchs, so the modern geologist calls out from their stony tombs the gaunt skeletons of Ichthyosauri, Pterodactyls, Mastodons, and Megatheriums, to astonish us with these relics of former creations. Geology has its being only in the past, but Astronomy exists in the present. The vast orbs which, in companionship with ourselves, roll round the sun; that mighty luminary itself, the centre and life of our system, exists now as surely as we do ourselves, and will exist in after-ages. The dominion of Geology is restricted to the past; Astronomy is with us in the present, and will with be us in the future.

Another and a very important distinction between these two sciences is a difference in the nature of the proofs upon which they rest. There are three great methods of inquiry by which we advance to the attainment of scientific truth, three roads on which the human mind travels in the acquisition of knowledge. The *first* of these is the study of the Abstract and Exact Sciences of Arithmetic and of pure Mathematics, whether in the form of Geometry or of algebraical analysis, and their subsequent application to the solution of all these problems capable of being submitted to them. The *second* is the accumulation, observation, comparison, and analysis of the Evidence of Facts, and the deduction from them of uniform laws of nature and fixed relations of cause and effect. The *third* is the system of Experimental Philosophy which is coupled with the great name of Bacon as its inventor. Some further consideration of the nature of these three methods may not be superfluous.

The first of these, viz., the study of abstract science, is distinguished from the other two by not being dependent upon the evidence of facts. When its principles have been determined, it is only necessary to ascertain that the facts are in accordance with the data required by the conditions of the problem; the result then follows as a matter of necessity. For example, we are perfectly sure that 2 and 2 make 4; and although we know nothing of the state of matters or of the conditions which exist in Sirius or the Pleiades, or any other of the constellations, yet we are perfectly certain that 2 and 2 must make 4 there as they make 4 here. We are sure that in no quarter of the universe can 2 and 2 make 3, or 5, or 7, or any number but 4; and it is not necessary to adduce facts in confirmation of this from any experience of the particular state of things in any of these stars.

What is true of this first and simplest in the series of the multiplication-table must be true of all the others, though they may not be quite so patent to the eyes; it is as certain that 7 times 7 make 49 as that 2 and 2 make 4, and the whole multiplication-table must be as correct in the star Arcturus as in any national school in the three kingdoms where it is painfully drubbed into the heads of puzzled ploughboys. The multiplication-table is one of the first stages in arithmetic; logarithms, are among the most refined and the most subtle, yet it is as impossible that the logarithm of the tangent of the angle of 44° should be any thing less than 9.984837, or that the logarithm of the number 365 should be any thing but 2.562293 as that 2 and 2 should not make 4. Where the principle is true the results must be always the same. Relying upon the correctness in the principles of logarithms, mathematicians use them to facilitate their most abstruse calculations. What is true in arithmetic is equally true in every description of pure mathematics. When the young student has mastered the *pons asinorum* in his Euclid, he knows absolutely that the two angles at the base of an isosceles triangle are equal; he does not require to seek for a confirmation of this in the evidence of fact; it is not necessary for him to measure 100 or 1,000 or 10,000 isosceles triangles in corroboration of the demonstration of the fifth proposition of Euclid; it is quite sufficient for him to have consulted the diagram and to have mastered the demonstration in order to convince him that, under all possible circumstances, and in every quarter of the Universe wherever an isosceles triangle is found, the angles at its base are equal. The demonstrations upon which rests the truth of Newton's "Principia," which may be regarded as the most refined examples from geometrical proofs, are as certain and as fully demonstrative as Euclid's fifth proposition can be; they are equally independent of the evidence of fact, and facts when in accordance with them in the planetary system must of necessity be bound by them. The same reasoning applies to all the methods of Algebraical analysis up to the Integral Calculus, and the great distinguishing superiority of this mode of reasoning follows throughout, that it is independent of the evidence of facts, but facts must obey its laws wherever the conditions are in conformity with them.

The second method of proof is by establishing an invariable and continuous connection of Cause and Effect, by showing that certain events invariably follow from certain antecedents, thereby establishing a necessary connection between them. By this method of reasoning and observation a very high degree of proof may be obtained, but it is not quite so overwhelmingly convincing as are the conclusions drawn from mathematical demonstration. Let us compare two examples of proofs obtained by each of these two methods. Nothing can be much more certain upon this earth than that every oak has sprung originally from an acorn, and that a sackful of acorns planted at the proper season and in favorable soil will produce a whole crop of young oaks; but our only ground for believing this is founded upon the past experience of similar facts. We can have no possible means of knowing, if shown an acorn for the first time, that it would produce an oak; no reasoning of the *a priori* kind could lead us to such a conclusion; we know that the consequence will be such merely from universal previous experience. The nature of the proof differs altogether from that by which we arrive at the certainty that the angles at the base of an isosceles triangle are equal. In the one case the result is certain under all circumstances, and arises from the nature of things. In the other it is a consequence which we have observed to follow under existing conditions from certain premises, but we cannot trace the connection ourselves, and only that, so far as our experience extends, it uniformly follows.

Mr. Babbage, in his "Passages from the Life of a Philosopher," gives an instance of the possible error which may lurk under these arguments, founded on the principle that similar causes will always produce similar effects. He says that he observed that he could set his calculating-machine to such a point that a certain combination would be produced a million of times, but at the million-and-first the result would be different. Any one, he said, merely reasoning from cause and effect would conclude at the 999,000th time that at the million-and-first time the same result would follow which had already preceded so many similar trials; yet he would be wrong, and his reasoning would be proved to be fallacious. Still, although the certainty is not so absolute or the proof so overwhelming as in cases where the reasoning is founded on mathematical demonstration, a degree of proof almost amounting to certainty may be attained through the evidence of facts. We must, however, admit the possibility of error, and adopt conclusions based upon the relations of cause and effect, with a slight reservation.

The third method of inquiry does not differ altogether in principle from the second, in the same manner as they both differ from the first. The second and third methods are both founded upon the evidence of fact and the relations traced between cause and effect. They differ from each other in the manner of attaining this evidence and of tracing these relations of cause and effect. In the second method the inquirer merely observes; he watches, and notes, and chronicles the operations of nature as he witnesses them; but in the process of experimental philosophy the enterprising and inquiring spirit of Man takes a bolder step. He is not satisfied with the mere passive part of an observer, he renders himself an active agent; he not only observes but directs, he brings into exercise that highest gift accorded by his Creator to Man, the faculty of imitating in however faint and imperfect a manner the acts of Omnipotence, and becoming in a remote and secondary degree an Intelligent

Cause. He is no longer satisfied merely to observe: he interrogates nature by the means of experiments, he asks questions of her, he twists the materials in his hands into new and artificial combinations, he brings substances together which would naturally have always remained apart, he probes the depths hitherto unrevealed. Sometimes he flings himself purely at hazard into the bosom of nature, and by accident discovers some novel secret; at another time he pursues some already discovered analogy up to fresh results; he resembles a skilful counsel cross-examining a reluctant witness, and he forces some concealed truth into light by dint of questions, each experiment being a question. By this process most of the great discoveries in the natural sciences have been arrived at. Experimental philosophy did not originate with Bacon, although the name of the system might. Those who first discovered that tin and copper combined produced bronze were experimental philosophers; those who discovered that a mass of grapes left to ferment produced wine were experimental philosophers, and very useful ones too. This third method of experimental philosophy is, in other words, the intellect and energy of Man guiding him to become an intelligent and active agent, producing those great discoveries by which he has obtained such mastery over the material world. It is the great element of that Progress which we hear so often spoken of and so frequently misunderstood; its power seems by no means exhausted, and it is quite possible that the energy of Man will add many new discoveries to those which have already rendered him the monarch of the material world. It is worth remark also, in days when a certain sect of philosophers seems always intent upon levelling Man to the condition of an animal, that this faculty of originating new effects by new combinations, the offspring of intelligence, is not shared in the remotest degree by any of the lower animals—it is the sole prerogative of Man.

These three methods of inquiry appear to me to be distinct from each other, but they are not incompatible; the philosopher may employ one or two, or all three of them, as he may find them to be most available for his purpose; in point of fact, in the study of most sciences they are so employed.

In applying these principles to the consideration of the evidence by which Astronomy and Geology are respectively supported, we find that Astronomy rests upon the double basis of observation and pure mathematical reasoning. The evidence of fact has accumulated in immense stores. The perfection which modern telescopes have attained, particularly since the days of the elder Herschel, multiplies a hundredfold the powers of the human eye.

Every appearance of the heavens, every motion of the bodies comprising our solar system, are scanned and noted with extraordinary accuracy. Within the last century two planets of first-class magnitude and importance (Uranus and Neptune) have been discovered. A numerous body of asteroids have been found occupying a large space in our system between Mars and Jupiter. The appearance of any comet within the circumference of our system is immediately perceived, and its course watched and measured. All the motions of the planetary bodies and of their satellites are distinctly visible to us through these great and powerful telescopes. Their changes of position and their revolutions round the sun can be traced with the minutest accuracy, and all this array of facts can thus be marshalled in confirmation of the truth of that great solar system which it is the glory of modern astronomy to have discovered. All this can be established by pursuing the second method—that of reasoning on the evidence of facts; but Astronomy can be supported on that still sounder and more certain basis of pure mathematical demonstration. The "Principia" of Newton and the "Analysis" of Laplace draw their proofs from the sources of pure mathematics. It is this combination of the results of these two methods which gives to Astronomy, at least so far as it relates to our own solar system, such absolute certainty in calculating the magnitudes, the distances, the velocities, and the motions of the different bodies composing the planetary system; we cannot conceive the possibility of error. The novel and bold theories of Prof. Stokes and others, founded upon the results of the spectroscope and calculations of the velocity of light, carry the speculations of the astronomer beyond the bounds of our own solar system, in an attempt to measure the distance of the fixed stars, and to trace motion in those bodies which hitherto were deemed stationary.

I will not attempt to follow those daring speculations through those new fields on which they invite us to enter. No doubt the enterprising energy of those philosophers will explore these new regions of science, should they prove accessible to the powers of Man. I am contented for the purposes of this essay to restrict myself within the bounds of our own solar system, which has been so thoroughly understood and established. Should it prove possible to future astronomers to extend our knowledge, no new discovery lying altogether beyond the limits of our own system can disturb the calculations we have arrived at, or shake the truths which have been so firmly established. We may attain a yet wider knowledge, we may complete our acquaintance of a still more comprehensive scheme of Omnipotence, but such additions cannot disturb the knowledge we have already attained. I will only remark, that the reasoning of these gentlemen seems in great degree to be founded upon the third method—that of experimental philosophy; the use of the spectroscope seems to belong to it.

When we turn to Geology, we perceive at once that it stands on a far narrower basis than its sister science; it derives no support from pure mathematics. A state of things which has passed away, and which we can only imperfectly conjecture, cannot afford the data necessary for mathematical demonstration. Perhaps some small exception may be made in favor of those calculations as to time, which are afforded by the accumulation of deposits at the mouths of great rivers, by the rate of the formation of peat morasses, or of the growth of primeval forests, which may afford a loose and general ground of arithmetical calculation. With these trifling exceptions, Geology must rest entirely upon the evidence of facts, and it must be owned that these facts are neither so numerous nor so well ascertained as those relating to Astronomy. As I think I already observed, they are derived from the relics of former conditions of our Globe; they cannot be tested by comparison with what now exists; they are also very much scattered and unconnected. We cannot refer to any former period of the world's history as indicating a state of things complete in itself; each of the various strata which geologists trace beneath the earth's surface seems to refer to a period of transition between the stratum immediately below and that above it. Each of these strata contains many of the older species of plants and animals, and each contains many new animals. At each step many of the older varieties disappear, and many new ones appear; but all this renders the evidence of facts upon which they are founded exceedingly complicated and difficult to trace. We long for the certainties of mathematical demonstration, and for the power of examining the present. Where all the theory is founded upon the evidence of facts, and of facts so scattered and so remote, our calculations are liable to be disturbed at every turn by new facts inconsistent with the results we imagined that we had realized.

An example of this may be found in the different theories respecting the age of Man and his first appearance upon the earth. The earlier geologists had adopted the conclusion that the creation of man corresponded in time very nearly with the chronology of Genesis; all their calculations have been disturbed by some later discoveries. There have been discovered under some peat bogs and morasses in the neighborhood of Amiens a great number of spear-heads rudely chiselled, but which are apparently the work of human beings. It is considered certain that a much longer period would be required for the formation of the morasses of peat under which these relics are buried, and this fact alone has shaken all the conclusions of former geologists. There have subsequently been discovered also caves in which human bones have been found in proximity with those of some species of quadrupeds which have long become extinct; and this has thrown still farther doubt upon the date to be assigned to the first appearance of man upon the Globe. No question can be more profoundly interesting to us, whether upon physiological, moral, or theological grounds, than that of the date of man's creation: and yet these latter discoveries appear to have enveloped it in the greatest uncertainty. Incomplete and defective as the evidence of facts in Geology appears to be, yet much has been established from the mine of facts, which is still rich in new materials; and the tendency must be to strengthen and improve our knowledge as fresh stores come to light. Many of the principles of Geology may be considered as established, and future labors will doubtless confirm them, and correct any existing errors.

The general conditions which geologists may consider as being established are, first of all, the immeasurable antiquity of the earth; secondly, the existence of different and separate periods indicated by various strata containing the fossil remains of animals and plants proper to them. The series of these strata is never found inverted; particular strata may be found wanting in any locality, but older strata are never found over more modern formations. It may be remarked that one great defect pervades Geology, as far as it has yet been explored, which is the entire vagueness and uncertainty of all its chronology. Whether the various epochs embraced 10,000 or 50,000 or 100,000, or 5,000,000 of years is a calculation on which Geology has hitherto failed to afford the slightest light, or to make any estimate.

The extent embraced by Geology is no doubt a very large one, particularly when considered with reference to the earth and to man. It includes not only the whole existing surface of the Globe, but a number of former surfaces which it would appear have been successively covered by new formations. Geology may thus be regarded as embracing not only the present surface of the Globe, but several other former surfaces now buried beneath it. When, however, we contrast it with the extent of the whole Solar System, it shrinks into comparatively small proportions. We find that the Earth itself is but a planet of a secondary magnitude and importance in the system. The eight principal planets which revolve round the Sun may be divided into two classes: the four nearest the Sun, consisting of Mercury, Venus, the Earth, and Mars; and the four most remote, composed of Jupiter, Saturn, Uranus, and Neptune. The difference in magnitude of the four nearest and the four most remote is very considerable, Jupiter being 590 and the Sun 602,000 times larger than the four Terrestrial planets combined. The division as to space also is great, since the first four planets revolve within an area the radius of which is not one-third of the distance between the Sun and Jupiter and one-twentieth of the distance to Neptune, the most remote of the known planets. The subjoined table of the diameters, volume or cubic contents, and the distances from the Sun of the eight principal planets, and of the Sun's diameter and volume, will render the comparison more obvious.

It will be observed, by the comparative estimate in this table (column 5), how small and insignificant is the Earth when compared with the larger planets, and, still more strikingly, with the Sun itself. There is one remark which suggests itself in this comparison, and that is, notwithstanding the immense disproportion in size between the Earth and the Sun, how much more close and intimate is their connection than that between the Earth and any of the other planets. If Jupi-

Name of Planet.	1	2	3	4	5
	Mean Diameter.		Mean Distance from Sun.	Volume, or Cubic Contents.	
	Miles.	Earth's=1	Miles.	Millions of Miles‡	Earth's=1
Mercury *	3,058	0.387	35,392,000	14,973	0.0577
Venus . *	7,510	0.949	66,134,000	221,778	0.855
Earth . *	7,912	1.000	91,430,000	259,333	1.000
Mars . *	4,363	0.551	139,311,000	43,486	0.168
Jupiter . †	84,846	10.724	475,692,000	319,811,264	1,233.205
Saturn . †	70,136	8.865	872,137,000	180,643,616	696.685
Uranus . †	33,247	4.202	1,753,869,000	19,242,302	74.199
Neptune. †	37,276	4.711	2,745,998,000	27,119,866	104.575
Sun . .	852,908	107.799	—	324,867,451,701	1,252,691.000

* The four nearer or terrestrial planets.

† The four farther or major planets.

‡ The cubic contents are calculated as if the planets were spheres, and no account is taken of their trifling spheriodicity.

ter should leave his place in our system, and set out on the grand tour through all the constellations of the firmament, or if Venus should elope with a comet, such events would occasion great amazement and consternation throughout all the observatories of Europe, scientific societies would be holding meetings in all the principal cities, and the gravest professors would be in a state of almost frenzied excitement. But all this while, if there were no newspapers or journals to tell the tale, it may be doubted whether the great bulk of the people would ever find out any thing about the matter. Neither Jupiter nor Venus would be missed from their accustomed places by ordinary mortals, nor is there any proof that their disappearance would sensibly affect the remaining members of the system; but we are all aware that light and warmth and existence itself depend entirely upon the presence of the Sun: were his rays withdrawn from us for even the briefest space of time, utter extinction would be the inevitable consequence.

Here I propose to limit my comparison to the Solar System. I will not venture to enter upon those larger speculations which have been lately opened by Prof. Stokes and others in the field of Sidereal Astronomy. I think it is sufficient for my purpose to point out the vast disproportion in size between the Earth and the other members of our own system, and show that our world itself is but a very small fraction of that great whole. When we endeavor to pass beyond the limits of our own system, we leave behind us our two great props and aids—the powers of the telescope and the resources afforded by pure mathematics. Before these speculations had been started, we possessed certain negative testimonies with regard to the sidereal system lying beyond our own. We knew that the range of our own telescopes extended to the utmost limits of our own system; as therefore they were powerless when applied to the fixed stars, it followed, as a natural consequence, that the latter were greatly more remote even than the most distant of our own planets. We might also infer that, as the light of the Sun can but feebly illuminate our own remoter planets, it would produce no effect upon those still vaster distances, and therefore that the fixed stars shine, like the Sun itself, by their own light, and are, in fact, also suns. Should the calculations of Prof. Stokes be confirmed, and the distances of the fixed stars ascertained, these discoveries would only confirm what we possess a negative knowledge of already—that these stars are infinitely remote. The same observation, however, would not apply to the fact, if established, that the stars are not fixed, but moved. Such a result would induce the belief, either that they wandered about in space without any guiding law, or that they were governed by some more comprehensive system of which we only could reach a part.

We now approach the consideration of an

essential difference of vast importance between these two sciences. The magnitude of the consequences involved in this difference it is impossible to estimate. The distinction upon which I am about to remark is no less than that of an absolute and entire difference between the main principles of astronomical and geological science. The Solar System is governed by a law of fixity; by which I mean, of course, not that the bodies which compose it are stationary, but that all their motions and the revolutions which they make round the Sun as a common centre are periodical, but are always the same. The Solar System may be compared to a vast clock, all the wheels and mechanism of which are perfect, and which keeps time with unvarying correctness. In illustration of this proposition, we may select at hazard any one of the parts of this great system, and the result will be the same.

For example, let us take the Earth's annual revolution in its orbit round the Sun. The distance traversed is about 574,310,000 miles, and the velocity at which the Earth moves in accomplishing this revolution is 65,518 miles an hour, or 18.2 miles per second. Yet this prodigious space is traversed by the

Earth every year to a second of time. The difference in the mode of computing the year, whether the sidereal or solar day is adopted, does not in the least affect this argument. The exactness of the Earth's revolution is, in point of time, the same in each. The consequences resulting from this are invariably the same, as is the motion of the Earth itself. For example, the lengthening and the shortening of the days in any given latitude are the same in each succeeding year. The motion of the Earth through its orbit is not less wonderfully exact when considered with reference to space than it is to time; it not only occupies the same time to a second, but it traverses exactly the same orbit. This may be verified by observations of the position of the Earth with reference to the constellations, and in passing through the different signs of the zodiac at different periods of the year. We find that its relative bearing to the fixed stars, which determines its place in the heavens, is always the same at the same period of the year.

Another proof of the absolute precision and uniformity by which all the motions of the heavenly bodies are regulated may be drawn from the methods to ascertain the longitude.

The general principle upon which the observations for this purpose are founded is of course familiar to all educated persons. The daily revolution of the Earth round its axis from west to east takes place in twenty-four hours. The 360°, therefore, into which every one of the parallels of latitude is divided, are traversed at the rate of 15° per hour, or 1° in four minutes. If, therefore, any particular meridian is adopted, say the meridian of Greenwich, as the line from which the distances in longitude are calculated, the difference in time between that meridian and any point on the Earth's surface will correspond with the distance in degrees and miles east or west of that meridian. Say, for instance, that when it is twelve o'clock at Greenwich, it is eleven o'clock at some point to the west of Greenwich, we then know that the distance is 15°; or if, on the contrary, it is one o'clock at the place of observation and it is twelve o'clock at Greenwich, we know that we must be 15° to the east of Greenwich. But, in order to institute this comparison, it is necessary to possess the means of ascertaining the exact time at Greenwich and at the place of observation simultaneously. The time at the place of observation is obtained by certain astronomical observations, which it is not necessary now to particularize. The time at Greenwich is calculated by two methods. The first and easiest is by a chronometer set to Greenwich time, and therefore indicating the hour at that place, wherever the chronometer is taken. The great improvements which have been made in the construction of chronometers has rendered this simple method one of considerable accuracy. It is exceedingly valuable to navigators in conjunction with that more scientific process, by lunar observation, as verifying a correction of any error in the reckoning, and also as filling up the blank occasioned by any interruption in the state of the weather. Pocket chronometers are also of the greatest use to travellers by land in the remote and unknown quarters of the globe, as in the interior of Africa or Australia, or in Central Asia, where a ready and portable means of determining the longitude of places is an invaluable aid to geography. Nevertheless, as the best chronometers are liable to irregularities and cannot always wholly be relied upon, the means of ascertaining the longitude by astronomical observation had long been a great desideratum to the navigator. The motion of a ship at sea threw considerable difficulties in the accomplishment of this object, where such great accuracy in the observation is requisite. A reward of £30,000 was long offered in the last century for the attainment of this object; but although a large portion of this reward was given to Mr. Harrison for his chronometer in 1767, yet, as perfect accuracy has never yet been obtained either by chronometers or, in consequence of the motion of the ship, by lunar observation, the whole £30,000 has never been adjudged. Still, however, the tables constructed by the Board of Longitude, and inserted in the Nautical Almanac, do give the means of determining the longitude with perfect accuracy, subject only to the error which may arise from the motion of the ship. It is evident from the foregoing that, as the determination of the longitude is arrived at by ascertaining the time at two places simultaneously, the astronomical means of effecting this is by converting the heavens into one great clock, from which the observer can read the hour as from a dial-plate; and in order to accomplish this, three things are necessary. First, to ascertain the exact time at the place of observation; secondly, to know, by the position of the Sun, Moon, and stars, the corresponding time at Greenwich; and, thirdly, that the motions of these bodies should be so uniform and regular as to enable the astronomer to calculate beforehand what will be their exact apparent distances from each other at every hour of the twenty-four. Now it is this last calculation which is made in the Nautical Almanac for every year, and which is in constant use in navigation, and which furnishes a convincing proof of two astronomical facts. First, the wonderful accuracy and regularity of the movements of the Earth and its attendant Moon throughout the whole of its annual revolution. Secondly, the thorough mastery which astronomical science must have obtained over the knowledge requisite for these calculations, so as to enable astronomers to calculate them beforehand with such perfect correctness. Let us see what are the elements upon which they are founded. The Sun and the stars may be considered as stationary, and the Earth as describing its annual revolution in its orbit through the void between the Sun and the stars; while this is being accomplished, the Moon is describing her monthly revolution round the Earth and accompanying it in its progress. Now it is evident that, although the positions of the

Sun and the stars are in fact stationary, yet they appear to alter to an observer upon the Earth as it proceeds on its course; but as the Earth pursues exactly the same track in each succeeding year, her relative apparent position to each of the stars will be the same as it was in the same day and hour in the preceding year; and also with regard to the Moon, which revolves round the Earth in the same manner, her apparent distances from the Sun and from the several stars will be the same; but the whole certainty of these results will depend upon the perfect accuracy of this complicated calculation, viz., the exact coincidence in time and place of the Earth's track round the Sun, and also of the Moon's revolution round the Earth. These data being given, and the position of the Sun and stars being assumed as fixed, the apparent distance of the Moon from different stars will be calculated with certainty. The calculations, however, are necessarily very complicated, as they involve the two motions of the Earth round the Sun, and of the Moon round the Earth, and when we see in the Nautical Almanac that these calculations are made to decimals of a second, we may form a conception both of the labor required and of the thorough knowledge of both these motions to calculate them beforehand with such minuteness.

A similar illustration of the uniformity in all the motions of the heavenly bodies composing our planetary system, and of the correctness with which astronomers are enabled to calculate them, may be furnished by the observations of the satellites of Jupiter. This planet, as may be seen in the foregoing table, has a mean distance from the centre of the Sun of 475,692,000 miles; its distance from the Earth varies in proportion to the position of each planet in their respective orbits. When the Sun is exactly between Jupiter and the Earth, their distance then is the whole distance of the Earth from the Sun, added to the whole distance between the Sun and Jupiter, viz., 567,122,000 miles; when the Earth is directly between the Sun and Jupiter, the distance is diminished by the distance of the Earth from the Sun, viz., 384,262,000 miles, and the distances vary within these limits, in proportion to the position of the planets in their respective orbits. Now at these immense distances, the moment of the eclipse of the satellites is calculated to seconds reduced to Greenwich time, and another means is thus given of ascertaining the longitude. Thus we find that the revolutions of the satellites of Jupiter round that planet are known to us with as much accuracy as the revolutions of our Moon round the Earth, and that they are equally regular and constant.

A still more convincing proof of the uniformity and regularity of the motions of the planets is to be found in the periodical transits of Venus over the Sun's disk. As Venus is 25,296,000 miles nearer to the Sun than the Earth is, the orbit in which she describes her revolution is contained entirely within ours; if, therefore, the orbits were in the same plane, or to use a more popular form of expression, upon the same level, Venus would be continually passing and repassing between the Sun and the Earth; but this does not occur, because the orbit of Venus is not in the same plane with that of the Earth, but is, as it were, tilted up on one side and down on the other, to use a very unscientific mode of expression; or, as astronomers would say, the orbit of Venus is inclined at a certain angle to the orbit of the Earth. When, therefore, Venus, in her revolution, passes between the Earth and the Sun, she does so, to use the same popular and unscientific mode of expression, either at a height above, or at a depth below, the exact line between the centres of the Sun and of the Earth; Venus only passes directly between the Earth and the Sun on those rare occasions when the planes of the two orbits intersect each other, so that a line drawn from the centre of the Earth to the centre of the Sun would pass through the centre of Venus. Now, in consequence of the angle of inclination of these orbits to each other, such a result happens but rarely; it only takes place at intervals of 105 and of 8 years, two transits following each other at the short interval, and then the long period of 105 years elapsing before another takes place. Now the proof of the great exactness and uniformity in the motion of both these planets is found in the fact that these transits can be foretold with the same precision as that with which we calculate the eclipses of our own Moon. This shows that in 105 years not the slightest change takes place in the revolutions of either.

Many other illustrations might be afforded in proof of the foregoing facts, establishing the perfect uniformity and exact periodical recurrence of all the motions of the bodies composing the Solar System, governed as they are by the universal law of gravitation. Indeed, there is no part of that system which does not corroborate the truth of these principles. Wherever slight disturbances have been discovered, they have been found to result from the effects of the law of gravitation acting by the different planets upon each other, and therefore slightly modifying the force of gravitation exercised by the Sun itself. These were first traced by Laplace, but they always furnished, upon examination, a corroboration of the proof of the great principles upon which the Solar System rests. The last instance of this kind is to be found in the simultaneous discovery of the planet Neptune by our own countryman Mr. Adams, and by M. Le Verrier, deduced from the observation of slight irregularities in the motion of the planet Uranus. But the calculations of astronomers prove that none of these irregularities permanently derange the system; they are compensated for in the whole action, and the stability of the system is never permanently affected by them.

The researches of astronomers have never found in any part of this glorious scheme the principle of decay; those who entertain a firm belief in the existence of the Great First Cause, acknowledge indeed with reverence and awe that He who created this great work of His omnipotence could, if He so pleased, destroy it; but there is no indication whatever to be found of such a purpose; as far as human knowledge and foresight extend, our Solar System bears the evidence of being eternal. We may remark that, as there is no perceptible symptom of the germ of decay, so, on the other hand, we can never discern any mark of progress; no fifth satellite is in process of formation to add to the light of Jupiter, nor is Mars furnished with a moon, although, his being more remote from the Sun than the Earth, he may consider himself as somewhat unfairly treated; the numerous asteroids, or small fragmentary planets, which float in the large vacant space between Mars and Jupiter, show no symptom of congregating together, although they might form a planet of respectable dimensions. Progress and decay are equally imperceptible in our Solar System.

Some theologists and divines mav be startled at this view, which they may consider at variance with the prophecies of Scripture, which so frequently refer to the end of the world; but I think that no such discrepancy will, on consideration, be found to exist. Both the Old and New Testament relate exclusively to the destinies of Man. Now whenever the Almighty should, in His wisdom, determine to end the mortal existence of man, such an exercise of His power would not be difficult to conceive. Life is so frail a possession that it would seem as if the exercise of the Divine power were far more needed to preserve than to extinguish it; it would but need the temporary abstraction of the oxygen from the atmosphere, or the escape of some mephitic vapor from the bowels of the Earth, to destroy, in an incredibly short space of time, all Life, whether human, animal, or vegetable; leaving the earth a *tabula rasa*, to be reoccupied by new forms, according to the will of the Creator. We may also remark that Geology shows us that various races ot animals existed at remote periods upon the Earth, and have become extinct from different causes: it is quite possible to conceive that analogous causes might extirpate man; but all these changes need not affect the course of the planets, or the position of the Earth as a member of the Solar System. Uninfluenced by these changes, the great mass of our Globe might continue to fill its accustomed place in the system of which it is a member.

The whole of the beautiful scheme of our

Solar System, so complicated and yet so perfect in its workings, gives the most convincing proof of design. Its vast mechanism is to be traced to the combined agency of two forces in nature directly opposed to each other, and therefore modifying the effects produced by each. The first of these is the universal law of gravity, by which all bodies are attracted to each other; the vast mass of the Sun would under this influence absorb the whole planetary system within itself, were not its action controlled by the centrifugal force imparted to all these bodies, and which gives them a tendency to fly off in a straight line, as a stone does when escaping from a sling. What that force is may be conceived, when we remember, as stated above, that the whole mass of our own Globe rushes through space with a velocity exceeding eighteen miles a second; and we may estimate the exquisite nicety of adjustment between these two opposing forces, when we consider that they are so accurately balanced as to produce, as I have stated above, the annual revolution of the Earth round the Sun to the fraction of a second of time. The same effects result from the same causes throughout the whole Solar System; everywhere we see the law of gravity controlling the centrifugal force, and compelling the planet to describe the elliptical curve marked out for it in space.

What are these two forces? whence do they spring? how are they governed? how modified so that their combined operation produces so harmonious an effect? We cannot tell; human reason only enables us to trace secondary causes to their origin in the First Great Cause. These secondary causes we call laws, but that term cannot explain their origin. The force of gravity is a law of nature which, as far as we know, is universally operative, but so is the law of centrifugal force, which is directly opposed to the other. What is the power which is superior to both these laws? which directs and moulds them and counteracts one by the other, and uses the two to accomplish a mighty design? There can be but one answer; the earliest instincts of the rudest nations prompt it; the loftiest flights of human reason can only enable us to repeat it.

Geology exhibits a strong contrast in principle to the Solar System, which I have endeavored to illustrate. It would be impossible indeed to conceive two principles more completely opposed to each other; throughout the whole Solar System fixity and uniformity of law is everywhere found, the vast mechanism works in obedience to a common principle, each part is consistent with every other, and no symptom of variation from the mighty plan can there be detected now, or anticipated in the future. The Earth, considered as one of the planets, and as far as it is associated in this common design, strictly obeys all its laws; indeed, I have drawn from the motion of the Earth and its relation to the Sun and the other planets some of the most striking illustrations of the principles I have sought to explain. Geology, however, does not contemplate the Earth in its relation to the other part of the system; its province is to consider the Earth in what may be termed its individual character, and to trace as far as is practicable the past history of the planet from the vestiges which are accessible to the researches of man. Viewed under this aspect, the leading principle of the Astronomical system is reversed; fixity seems to govern the one, change appears to guide the other. When we follow the Geologist in his explorations, and penetrate those strata from which he extracts his evidence of former conditions of the surface, one truth appears to be evident—

that the Earth has undergone a series of transformations. Whatever uncertainty may prevail as to the date and manner or the cause of these changes, the fact seems to be indisputable. Primitive strata are found at the bottom of the series in which no trace either of animal or vegetable life can be discovered; over them are found more recent formations, mounting in an ascending scale to the animal and vegetable productions, which are preserved in them in a fossil state, indicating a closer resemblance to our condition.

Different Geologists have assigned these changes to different causes. Some, particularly the earlier ones, have adopted what is termed the theory of catastrophes, and have supposed the crust of the Earth broken by violent internal convulsions, lofty ranges of mountains thrown up, volcanic action modifying the whole surface of the Globe, earthquakes and deluges transforming the face of nature. Others have attributed more powerful effects to quiet and imperceptible agencies constantly operating, such as the detritus of the higher mountains carried down by torrents or by rain, the alteration in the level of the sea occasioned by deposits at the mouths of large rivers or by currents operating throughout its extent. Some have adopted the igneous theory, and have supposed that the Earth was originally launched forth in an incandescent state, and has gradually cooled in the lapse of ages. Others have taken up the glacial hypothesis, and have found traces of a period when the Earth was much colder than at present and covered with snow and ice.

All these theories, which it is difficult indeed to prove, may have some foundation. If we assign to the World a vast duration in point of time, it may have passed through successive periods containing all these diversities, and each leaving its mark. Traces of all these agents may still be found, and some are still actually at work. Earthquakes and volcanoes are not extinct, and the sea is constantly receiving the deposits brought by rivers into its beds.

In following the footsteps of Buckland and Murchison, Owen, Cuvier, or Lyell, and the many other distinguished Geologists whose names will occur to us, we reverence the intellect and sagacity which in the course of less than a century has opened a new field of science to our inquiries. Yet in pursuing the tracks left by these enlightened minds we are often tempted to regret the loss of those two great aids to Astronomy—pure Mathematics and the assistance of the Telescope, and the observation of bodies actually existing before our eyes. The Geologist gropes comparatively in the dark; and the greatest caution is necessary before we adopt conclusions founded upon the evidence of facts alone, and of facts so buried in the depths of time. How much has perished! how small a residuum has escaped the ravages of the Great Destroyer upon which to rear our geological structure! Yet how careful ought the true philosopher to be in substituting hypothesis, however plausible, for the sole proof he can adduce—the evidence of facts—or in supplying by the same means the hiatus so often left in the chain. Where our resources and means of proof are so slender and limited, where the facts are so few and so scattered, how easy it is to adopt some erroneous conclusions and to follow some false lead. The bases upon which Geologists seem to be agreed are: 1. The great, though unknown and incalculable, antiquity of the Earth. 2. The series of changes which its surface has undergone. 3. The long period of time which must have elapsed while the Earth was passing through these stages; the record which they have successfully left in the strata which they have accumulated, and in the various deposits which have been preserved embedded in each of these strata; the regularity of the succession in which the strata occur, showing that the periods were the same throughout the Globe.

This portion of Geological study has reference only to inanimate nature; but by far the most interesting and important branch of it relates to those vestiges and relics of Life which are sealed up in these strata, the great repositories of the past. Here it may be observed the great interest in Geology centres, and in this branch of inquiry it has an undoubted superiority over its sister science. Mighty as are the prospects revealed to our wondering eyes by Astronomy, there is a solitude in the spectacle offered to us: one great Living Spirit indeed breathes through the whole the divine intelligence of the Great Creator of the Universe; His infinite power, wisdom, and goodness, are displayed upon their largest scale. But Astronomy reveals to us no secondary or subordinate intelligences, animating with the breath of Life those vast material organizations. It is indeed highly probable that such exist, and that

Life in some shape or other pervades those vast habitations so well calculated for its reception. We see that light and heat are everywhere diffused; the planets have their atmosphere of air around them; the alternations of day and night, and of spring, summer, autumn, and winter, are found in them. All these analogies lead us to believe that Life, which is evidently so important a feature here, and which is so bound up in the existence of our own planet, cannot be absent from the other members of the system. Nor is it probable that the great Sun itself, the centre of all, the source of life here, and so far exceeding all the planets in magnitude, should be destitute of that vitality which he diffuses around. Certainly it would be as easy for Omnipotence to create living beings in the very bosom of fire as to enable fishes to exist in the element of water which is incompatible with the life of terrestrial animals, and to adapt their organs in turn to an existence on land which is cut short by a brief immersion in the other element. The author of the "Plurality of Worlds," a work attributed to the late Dr. Whewell, controverts this view, and labors to demonstrate that all the other planets are untenanted, and that life only exists with us. I have always considered his reasons as quite unsatisfactory, and, although nothing upon the subject can be proved, yet that all the presumptions are in favor of a contrary hypothesis; still, however, we must admit that nothing on this subject can be known with certainty; the condition of the planets must from their remoteness be utterly unknown to us; no telescope can do more than reflect the general form to our eyes, and even our own Moon, comparatively so near us, only exhibits a variety of lights and shadows upon her surface.

That portion of Geological research which is dedicated to the study of the remains of vegetable and animal Life is by far the most interesting and important branch of that science. This will appear very distinct to us if we picture to ourselves Geology deprived of this part of its investigations and reduced to a mere examination of the different strata of rocks and soil which have been successively accumulated. What interest should we feel in knowing that the lowest stratum consisted of primitive granite, or that others were composed of old and new sandstone? Take away the remains of those living creatures which have once animated the Earth, in these its stages, and Geology would then became a dull and profitless study indeed. The difference between granite and trap, or trap and sandstone, would become very unimportant in our eyes, if these formations did not tell us a tale of the former living inhabitants of the Globe. These investigations also lead us to a conclusion of the most momentous character: we find, in tracing the Zoology of the Globe, that a certain chain of succession is established, and that the principle of Progress appears to have pervaded it. Here is a broad distinction between Astronomy and Geology. As I have already observed. the stupendous mechanism of Astronomy evinces no germ of decay and no symptom of progress. In every stage Geology exhibits both. Without those traces and remains of former life Geology would resemble a theatre without actors or *dramatis personæ.* It is by this chain of events that the past is linked to us. There is another feature which distinguishes Geology from Astronomy under this aspect. There is not one tittle of evidence in Geological science to show that the changes it exhibits have any thing of a cyclical or periodically recurrent character. Ichthyosauri or Plesiosauri, Mastodons and Megatheriums, have disappeared from the Earth forever; there is not the slightest probability that we shall ever revert to the state of things in which they had their being. Not only does one organization gradually supplant and efface a preceding one, but viewed as a whole there is a constant advance in the structure of the animal life, the links of which are thus handed down to us. Some of the earlier formations are composed of conglomerations of shell-fish, the very lowest forms of life of which we have cognizance. What an immense advance is indicated in the fossil remains of the animals of a later date, showing that they like ourselves were furnished with the organs of sightand of hearing, which continue to be after so many ages the great avenues of perception. These senses mark a vast increase in the faculties of the beings who possess them. In order to make use of these senses these animals must have been endowed with instinct, that mysterious faculty, approaching so nearly to reason, and yet so plainly distinguishable from it.

Change does not necessarily include Progress, but in this case both change and progress are manifested. The word "progress" may be coupled with others giving it a signification of deterioration, as the progress of decay, the progress of ruin, the progress of destruction; but when the word progress is used by itself I think that it is popularly coupled with the idea of advance. We mean progress in good, progress from a comparatively worse state of things to a better; at any rate, it is in this sense alone that it is applicable to Geology. As far as we can trace the history of the past this idea is impressed upon us; it is a regular series of advancing steps. In order to measure progress, it is requisite that we should adopt some point of departure or comparison. A ship may be proceeding rapidly on its voyage, but a passenger in the cabin has no means of testing the motion. We are empowered in some degree to measure the progress which Geology enables us to realize, because we can estimate the point from which it departed, and we know that which it has reached. If there be truth in the discoveries which have been made, the earliest state of the Earth was entirely barren both of vegetable or animal life. The present condition of the Earth in this respect is known to us; every portion of land, every depth of the ocean is tenanted by swarms of living beings, each jostling its neighbor, and the amount of life is only limited by the quantity of aliment necessary for its subsistence. Above all these creations of different epochs towers Man himself, so incontestably the monarch of the visible World.

Now mere change may be the offspring of chance. The change accomplishing ends and directed through ages to purposes, dark perhaps in their origin, but which are now revealed to us as having been accomplished, such change and such progress prove intelligence and design. Under this aspect the Will of the Great First Cause is as apparent through all the mutations which the Earth has undergone as it is in the fixity which reigns in the Solar System; only it may be inferred that, as the modes of action are so dissimilar, the objects to be attained are different. Geology displays the history and course of Life, from its earliest birth to its present stage of development. If we ask, what is Life? what its principle? or what is its essence? we shall soon discover that we have approached one of those boundaries which the human faculties in their present state are unable to overstep. There are certain primary elements into which all our inquiries resolve themselves, and by which our most subtile reasonings must be limited. There are some exceedingly just observations on this subject in Mr. Buckle's "History of Civilization," in which he renounces the attempt to explain or to understand the nature of the Soul, and he shows very clearly that we do not possess the data upon which to found any practical result. After two centuries he indorses the truth of the couplet that those who pursue those vain and fruitless studies are destined

"on metaphysic ground to prance,
To show their paces, not one step advance."

Life is another of those great mysteries very much akin to this, but it is curious that we have the most perfect idea of the meaning of the word, although we cannot explain the nature of the essence. When we speak of animal, vegetable, or human life, we know perfectly well what we mean by it. The same remark applies to others of these simple and primary ideas; the difference between instinct and reasoning, for example, is perfectly intelligible to us in their effects; we have no difficulty in distinguishing their operations from each other, although we cannot trace or explain the nature of this difference from its sources.

I will not attempt to trace in detail the

different stages in Geology through which progress is marked. It is far from my purpose, as it certainly would be beyond my ability, to write a work on the principles of the science, or even to attempt a manual or elementary treatise upon it. I will content myself with referring to the admirable works which have been produced during the present century. Geological discoveries have required immense labor and research, and vast knowledge in their authors, but they principally rest upon unwearied diligence and mastery of detail in the incessant comparison upon which they are founded. They are not based in the same degree as Astronomy is upon abstruse Mathematics, requiring long study of abstract principles. It is not so difficult for us to follow and in a general degree to comprehend the outlines of the Science when they have been traced for us by the sagacity and intellect of these great originators.

Geology may be divided, therefore, into these two parts, the *first* relating to the material structure of the Earth itself and the various revolutions which it has undergone, and *secondly* the different phases of animal and vegetable life which seem to have accompanied or followed these physical and material changes. As I observed above, it is one of the imperfections of this Science that its chronology is utterly vague: we do not seem to have the slightest means of calculating the duration of time in the successive epochs of the Earth's surface. All that the Geologist hopes to demonstrate is that these changes take place in a certain regular succession, which he has been enabled to classify, and that the animal and vegetable creations correspond in a considerable degree with these epochs. They do not, however, appear to do so strictly: the animal and vegetable productions of which the relics are found in the different strata are not separated from each other by a sharp line of demarcation coincident with the strata. It would appear that the animal and vegetable productions of an earlier age gradually fade and melt away and are replaced by a gradual and diversified process by those of a later, the older ones rarely becoming abruptly extinct, but some variety of them being long retained, generally in modified forms. Particular periods, however, seem to be marked by new changes, the full value of which we are enabled to estimate by the light of our later anatomical knowledge: thus the formation of the eye and ear must have created a new revolution in animal life, and the creations of vertebrated animals with their apparatus for communicating sensation from the head to every part of the bodily frame must have constituted another great step in the progress of animal life.

The spectacle which Geology presents to us when we survey this general retrospect is a very sublime one; it widens to a degree almost equal to that of Astronomy our conception of the vastness of the scheme of Omnipotence. During lapses of time which we are utterly unable to measure, and under alterations of circumstances and conditions putting an entirely new face upon nature, we still trace one great design never lost, gradually making its way onward and working to the surface in each succeeding transformation. If we venture to draw a conclusion from this view, it is that Life is the principal object of a Creator's care, and that the material framework in which it is set is subordinate to the purpose of its gradual development and ultimate perfection. From the earliest and faintest traces of lichen upon the rock to the actual state of the World there is a continuity which includes all the links in one vast whole. We are, however, constantly reminded that this is a very new science: it indefinitely extends the area of our conception, but it rarely traces bounds or limits. If we take, for instance, Sir Charles Lyell's able work on the "Antiquity of Man," he has succeeded in unsettling all our previous ideas, but he has not substituted any others of a fixed character; he does not the least inform us whether Man has existed 10,000 or 100,000 years upon the face of the Globe. Perhaps this imperfection will be owing to the recent birth of this branch of knowledge, and the labors of future Geologists may give us additional clews through the labyrinth of the past.

In pursuing that comparison between Astronomy and Geology which it has been the object of this treatise to institute, we must conclude that the advantage in point of certainty largely preponderates in favor of Astronomy. The retrospect to which Geology invites us is indeed a noble one in suggesting the grandest speculations upon the birth and youth of the Creation. But Astronomy, still grander in its proportions, inasmuch as it deals not only with this World, but with the Universe, is, as far as our own mighty Solar System extends, perfectly clear and definite. There is no vagueness in the magnitude of the Sun or of the distance of Neptune; all is thus far as precise and accurate as it is elevating.

REMARKS ON THE THEORIES OF MR. DARWIN AND MR. BUCKLE.

I CANNOT take leave of this branch of the subject, more particularly that portion of it which relates to the fossil remains of animals, and the progress of organized life indicated by Geology, without some reference to the theories of Mr. Darwin, which have attracted so large a portion of the attention of men of science. If I were to omit all mention of works which tread the same ground which I have been traversing, I should appear to evade all the questions which he has raised, and which have engaged the attention of Natural Philosophers for several years past. The doctrines which he has given to the World are in many respects irreconcilable with those views which I have ventured to embody in the preceding pages. Should I leave them altogether unanswered, I might as well abandon entirely my own opinions as having been proved untenable. I feel very reluctant in a treatise so slight as this to engage in any controversy with so eminent an authority, particularly when I feel that a cause so momentous as that which I wish to advocate may sustain injury by my inability to render it adequate justice. No one can fail to recognize the scientific and literary merits of Mr. Darwin, no one can be blind to the extent of his knowledge or to his intimate acquaintance with the whole subject of Zoology. It would be the highest presumption in me to endeavor for a moment to question his authority upon any of those subjects falling within the range of the studies upon which he is an oracle, yet, while I am glad to receive instruction from him where he pours forth the ample stores of his information on all the subjects of Natural History, I may be permitted to examine the reasoning which he founds upon them, and to remark upon those points in which, as it appears to me, he fails to establish his principles from his facts. Mr. Darwin has all the zeal which naturally belongs to the inventor of a novel doctrine, and it is just to him at the same time to observe that it is quite free from any mixture of dogmatism or intolerance of opposite views; he has a profound conviction of the soundness of his own theories, and invites all adverse criticism with philosophical calmness and with a strong confidence in the justice of his own system. It is due both to his high authority and to the tone in which he places his doctrines before us to consider them in the same spirit of impartiality and moderation. He invokes the judgment of our reason, and even should that judgment be adverse to all our preconceived opinions he has a right to claim it.

The march of true Science is irresistible; wherever her foot is once planted it can never be withdrawn—she is the mightiest of conquerors. Every thing must give way before her; however unwilling or reluctant mankind may be, they have no choice but to follow wherever she has once taken an authoritative lead. Just as bows and arrows have given place to muskets, and brown bess to breech-loaders, the old sailing vessels to screw-steamers, and the screw-steamers to iron-clads, so in every other path, whenever it is once clearly and fully illumined by the torch of Science, the human race have no choice but to follow. If fifty years ago any speculative chemist had foretold the possibility that by a combination of Chemistry and Optics it would become possible to fix the fleeting images and shadows that flit before our eyes and transfer them to paper in shape of durable pictures, he would have been regarded as a dreamer. Yet the art of Photography has within that period been diffused over the whole of the civilized world, and in the future has become as imperishable as civilization itself, with which it is entwined. A still more recent and far grander and more important step in Science has been made by the Electric Telegraph. In the commencement of the present century the wonders it accomplishes would have been considered on a par with the inventions in the "Arabian Nights' Entertainments;" we should have regarded as the most incredible of fictions an Assembly in London transmitting a message to Calcutta, and receiving an answer in the course of the same evening before it separated, or the speech of the President of the United States being circulated in London the day after it was delivered. Such apparent miracles would have been ranked with the feat of the African magician removing Aladdin's palace in one night from the centre of Asia to that of Africa. Yet the discovery of the almost instantaneous transmission of the electric fluid once made, and the means devised of utilizing it having become a permanent addition to human knowledge, it will be preserved and extended by its own utility. No conceivable revolution in society can ever again wrest it from us; it must remain a permanent addition to the powers of Man.

The strongest example, however, which can be adduced of the permanent nature of every solid advance in Science may be drawn from the establishment of that Solar System from which I have already drawn several illustrations. It would be impossible ever to revert to the erroneous systems of Pythagoras or Ptolemy. The Earth can never again be enthroned as the centre of the Universe, but must be contented through all time to occupy her assigned place whirling about the vast orb of the Sun, to which we bear about the same proportion as a cricket-ball to the dome of St. Paul's.

The same observation applies to every discovery which human invention has made in the domain of Science; each, when once proved to rest upon truth, is in its nature indestructible. It is fortunate that these victories of Science have always proved to be blessings, and it is one presumption in favor of the truth of any new discovery that it is beneficial in its consequences. Our experience does not yet afford us one instance of any fresh stride fully made and established in the whole circle of human knowledge which has not tended to the improvement and to the moral and intellectual elevation of Man : whenever any novel theory may be broached which leads to narrow our prospects or to lower our position, the strong presumption is created that it is founded in error.

Where, however, the principles enunciated are so novel and startling and their consequences would give such a shock to all our most cherished convictions, as in the case of Mr. Darwin, it is natural that we should scan them strictly, and examine them in a spirit of doubt, and almost hostility. We must be aware of the tendency of minds, even of the most powerful and original caste, to become the dupes of their own inventive genius, and to construct from very slender materials the most plausible systems. Descartes was one of the greatest intellects of modern times, yet his whirlpools were among the most notable instances of those false lights which lure men of science into the path of error.

I will not venture to follow Mr. Darwin through the vast field of Natural History which he treads with such a wonderful variety of knowledge. Independently, indeed, of the conclusions to which he would lead us, his works afford an interesting study to the general reader. Dr. Johnson remarked of Oliver Goldsmith that he was writing a work upon Natural History, and that he could make it as interesting as a Persian tale. A similar remark may be applied to Mr. Darwin with still greater force, while handling his subject with far greater ability than Goldsmith could ever attain to. There are, however, certain leading principles embodied in his writings, which are, as it were, the pivots upon which his system is founded, and upon these I shall offer some remarks. They are the parts of his reasoning upon which he has failed to carry conviction to my mind. Throughout his works it appears to me that he ignores, or, perhaps we may be justified in saying, denies, all evidence of the operation of the First Cause: his whole argument is founded upon the endeavor to trace the entire scheme of Nature to the operation of secondary causes only, which seems to be set in motion, according to his views, by some self-acting process. It would seem that Mr. Darwin rejects altogether the belief in any direct influence of a higher power —to use his expression, "analogous to though superior to that of human reason." Now I do not wish to apply to Mr. Darwin hard names, or to carry his principles beyond any limits within which he himself desires to confine them. I cannot, however, affix any other term to these doctrines than that they contain a profession of pure atheism: the words "atheism" or "atheist" do not occur in Mr. Darwin's works, but the idea seems to me to be clearly presented to the mind. In the passage which I have just quoted, he seems to reject altogether all belief in the agency of a higher power, "superior but analogous to" our own faculties of human reason and will; yet it is precisely this conception which can alone enable the human mind to attain any idea of the nature of God.

Man's faculties are limited, and, however wide and comprehensive may be their range, he cannot form notions of which the materials are quite beyond his knowledge or experience. It is only by looking inward, and forming an ideal conception of a Being possessing those qualities in their highest perfection, which man only possesses in a very incomplete and defective degree, that we are at all enabled to realize to ourselves the nature of the Almighty. It is perfectly true that the Deity may possess other attributes than those with which we can invest Him, and which are as utterly incomprehensible to us as the idea of color might be to one born blind; but these qualities we cannot possibly know or comprehend. Let us take the idea of Wisdom, for instance: the quality does not exist in any one of the brute creation. When we speak of the wisdom ot an elephant, we only do so in a figurative sense; it is in Man alone that such a faculty can be said to exist. In order to form a clear conception of the meaning of Wisdom, we must embody it in some human form before we can reach it in its abstract signification; we must think of the wisdom of Bacon or ot Newton. In the same manner the idea of Power (I mean moral and material power combined) can only be embodied in the case of some human being, as the prophet Mohammed or the first Napoleon; and in like manner Goodness or Virtue can only be personified in Man: no animal can possess moral excellence. It is only by forming a conception of a Being combining these three attributes in their highest perfection, that the human mind is enabled to attain any comprehension of the nature of God. Such an idea may be very imperfect and incomplete, far below the majesty and goodness of the Supreme Being, but the utmost stretch of our faculties can only enable us to reach so far. It is only by looking into ourselves, limited and finite as the prospect must be, that we can find any material wherewith to enable us to realize, even in the faintest degree, the idea of the Deity. This is exemplified in the fact that when we describe the action of the Deity, we can only do so by availing ourselves of terms borrowed from our own senses or faculties. We speak of the all-seeing eye of Providence, or we say that the ear of Heaven is ever open to the prayers of Man. We do not mean by these expressions to convey a belief that the Almighty really makes use of such organs; we cannot possibly conjecture what are His means of communication with us; we only use those words which convey to our own minds the nearest approximation to the idea of His Omniscience.

When, therefore, Mr. Darwin peremptorily rejects, in his scheme for the production of Life, all agency of a higher power, analogous however superior to that of Man, he does, in fact, deny the agency of God, for he denies it in the sole form in which Man could be able to recognize it. Now how awful!

how tremendous a doctrine is this! The mind recoils from the consequences which would result from its adoption.

There is a method of reasoning not unfrequent in Mathematics whereby any proposition is proved to be false by assuming hypothetically that it is true, and then showing that the conclusions it leads to are absurd. Such a mode of reasoning may be applied to Mr. Darwin's system, which repudiates, if not by name, yet in fact, all the operation of an Intelligent First Cause, and traces all advance to what he calls the Evolution or spontaneous action in the bodies themselves. I do not know how otherwise to interpret his theories; for if he does assign any place to the influence of an external power, or the promptings of an invisible Intelligence, then he does exactly that which he refuses to admit when he rejects the action of a force superior but analogous to the intelligent will of Man. I therefore believe that I am not misconceiving or misrepresenting Mr. Darwin when I assume that in his theory of Life he regards it as affecting all the transformations it presents, and generating all the varieties it exhibits, by the working of certain occult principles which it generates by an insensible and involuntary action, as I translate his two main agencies—Natural Selection and Sexual Selection.

In his summary to the work on the "Origin of Species" Mr. Darwin seems to admit that all Creation springs from four or five primordial forms, but he observes that a more correct deduction from his principles would lead to the conclusion that every thing originated from one prototype. Mr. Darwin does not very clearly explain how he understands that this first step was made. He quotes the opinion of a "celebrated author and divine" who, he says, has written to him that—

"he has gradually learned to see that it is just as noble a conception of the Deity to believe that He created a few original forms capable of self-development into other and needful forms, as to believe that He required a fresh act of creation to supply the voids caused by the action of His laws."

Does Mr. Darwin adopt this theory of his correspondent? If he does so, then he admits the existence of an Intelligent and Omnipotent First Cause, only that the mode of action is not direct but derived through a chain of successive causes and effects. If, according to the correspondent's idea, he conceives that the Deity created these four or five primordial forms of animal and vegetable life, and that the successive stages of development by which all animated nature has attained its present varied shape necessarily followed, then his correspondent or Mr. Darwin himself must believe one of two things: either that the Deity created these four or five primordial forms, ignorant of the consequences which would result from them, or that He did so cognizant of the germs that they enclosed and foreseeing or predestining their ultimate results. But if the latter conclusion is adopted, then all the works of Creation follow from an Almighty and Omniscient First Cause just as much and as completely as if they had emanated more directly and immediately from His hand.

A clock or watch maker who makes a watch is doubtless an ingenious mechanician, but one who could contrive a machine which by turning a handle would produce watches as complete as those made by hand, would doubtless be a far more skilful mechanist than the first; but the watches made by the machine would be quite as much the work of the inventor as those made by his hand alone; they would equally be the products of his inventive faculties, and the only difference would be that it would require a higher order of inventive faculty and more elaborate execution in the case of the watches produced by the action of the machine.

It would be satisfactory if we knew with greater precision to which hypothesis Mr. Darwin himself adheres: whether that of his correspondent, or the one which appears to follow from his own explanations. The difference is very wide: in the one case he admits the whole scheme of life to be the creation of an All-wise and Omnipotent First Cause, although he places that First Cause at a distance infinitely remote; in the latter case he ignores the existence of a First Cause altogether. Supposing that Mr. Darwin admits the first hypothesis, viz., that of his correspondent, what is gained by remitting the action of the Great First Cause to so infinitely remote a period? If He purposed and designed this long series of causes and effects from the very beginning, He is in fact just as much the author of the whole as if He had resorted continually to a more immediate and direct exercise of His Power and Will. Why conclude that He never has exercised and never does exercise that Power and Will directly through all those countless ages? We pay a ready tribute of admiration to the vast amount of Mr. Darwin's knowledge, ranging through the whole circle of the Natural Sciences; the amount of his materials is overwhelming; still I cannot, for one, accord my conviction to the truth of the complex system into which he has fashioned them.

As I mentioned in a previous part of this Essay, all Science is founded either upon abstract mathematical reasoning, upon the evidence of collected facts, or upon facts elicited by experiment. Mr. Darwin's theory is founded chiefly upon the two first: his hypothesis of Natural Selection rests in the first place upon the mathematical and arithmetical truth that as all forms of Life are preserved or maintained by some sort of aliment, and that as the food, if not a finite quantity, at any rate increases at a much slower rate than Life would, if supplied with food to an unlimited extent, therefore the amount of Life is necessarily confined to the amount of food. This, as Mr. Darwin observes, is the doctrine of Mr. Malthus (which he regarded solely with relation to the human race) extended to all living beings. Mr. Darwin draws this conclusion, that the struggle for food, which is in other words the Struggle for Existence, to which food is necessary, pervades all animated nature; to this extent we must admit the soundness of Mr. Darwin's principles; organized life is everywhere dependent upon food, and the chief business of existence is to satisfy the want of it. But Mr. Darwin's next step in his chain of reasoning appears to be far more doubtful; he traces the creation of all the diversified forms of Life to the results of the competition for food universally going on throughout the World. This competition he conceives crushes the weaker, gives advantage to the stronger, and as he supposes that small varieties begin in the production of different forms of Life, that these small varieties where they tend to the advantage of the being are gradually developed by hereditary succession, by sexual selection, or by the favoring conditions of climate and other causes, into permanently different species. Now does this theory rest upon the evidence of facts sufficient to prove it, or are his facts eked out by a vast number of hypotheses which he cannot prove? It is very remarkable how often, in almost every page of his work, he makes use of the verb "to suppose." He is always asking us to suppose something which explains or supports his reasoning, but we are not disposed to admit these hypotheses (as facts) built upon suppositions—in so very intricate a labyrinth as he calls upon us to enter. He finds it impossible to construct his system upon the materials afforded by the actually existing World: he must press into the service all the antecedent Geological periods which that Science has enumerated, for he requires countless ages in order to give him time enough to work out the slow results of Natural Selection. He speaks of millions of years, and in his diagram he represents 10,000 or 14,000 generations; but the existing World, of which either from history or tradition we have any knowledge, is comprised within at most 4,000 years. A generation is perhaps rather an indefinite term: we may easily conceive a father having a son at 75 years of age who may live to be 75, and these two generations would amount to 150 years; but if we take the average duration of human life and the succession of one generation to another, 100 years might be calculated as numbering three generations; therefore the whole number of existing generations of Man, of whom we can have the faintest knowledge have existed, is 120. It is quite evident, therefore, that Mr. Darwin assumes as the basis of his argument periods of time purely imaginary;

he can have no shadow of proof that the World has existed for 14,000 generations, or 470,000 years. If we were to admit for the sake of argument that the solid sphere of the Earth might then have been in existence, how can we feel sure that that succession of Life necessary to his theory can have been preserved through such an incalculable period of time? Geologists tell us that the earliest formation is of primitive rock, without any sign of animal or vegetable Life; in the course of 470,000 years three or four of these periods may have existed, and every vestige of Life may have been swept from the Globe, and creations have begun again three or four times in the interval. Mr. Darwin requires for the development of his theory enormous periods of time, far exceeding any of which we have the slightest knowledge; this alone places his whole system beyond the domain of fact and in the regions of mere reverie and imagination. It is also necessary that there should be a continuity preserved during the whole period; the same series must be preserved during the whole 470,000 years, for if at any period a total destruction of Life had occurred, it is quite evident that this process of Natural Selection would have to be begun *de novo*. It is a weak point in Geological Science that its chronology is absolutely vague. No one can assign any limit of years to the duration of any one of those periods into which Geologists have divided it, and all argument therefore based upon any assumptions with regard to time must be absolutely worthless. I cannot feel at all convinced of the fact that, even on Mr. Darwin's own hypothesis, and granting to him all that he asks us to admit, all the operations of nature can be effected by this admittedly tedious process of Natural Selection. We must remember that this term Natural Selection does not really express any exercise of the will or of the understanding; it is a purely involuntary course upon which the living being, whether animal or vegetable, is forced by the pressure of circumstances. There is no guiding direction from above; an accidental variation is supposed to confer some advantage in the struggle for existence, and the advantage thus obtained secures for itself a certain stability in the future, which gradually ripens the accidental variation into a permanent one, and the permanent variation into a species. In all this there is no exercise of choice or of will, no intelligence exerted either by the being himself or on the part of any Higher Power. Results which we have always supposed to display an Intelligent Will either proceeding from the individual being himself or from a Higher Power, or from both, are under Mr. Darwin's theory the consequences of these different directions into which these bodies are hurried by the forces to which in this struggle for existence they become exposed. Now it appears inconceivable that many of the operations in nature can have been effected, or many parts of that great scheme which we see in action can have been framed, by this process of Natural Selection alone. We must recollect that on Mr. Darwin's principles foresight is excluded. Foresight is an attribute of Intelligence; if any being was supposed to exist gifted with the power of previously knowing and causing these results, then they would cease to be due to Natural Selection alone. Now even allowing any amount of time for the operation, there are very many of the works of nature which I cannot conceive could possibly be accomplished without the agency of Intelligence. For instance, one of the principal means through which Mr. Darwin supposes that changes are wrought is the agency of Sexual Selection; but in order that this power should be exercised, it is necessary that these living beings should be divided into two sexes; Mr. Darwin tells us that in fact a vast proportion both of the animal, insect, and vegetable kingdoms, are so divided.

The propagation of the species is the consequence of sexual connection, which requires an elaborate adaptation of the male and female organs to each other. Now how, in the first instance, could this division into the two sexes have been effected by Natural Selection? Did it occur in the parent stock, and was it handed down through the successive varieties, or was Natural Selection to effect it again in each particular case? Supposing that the first was a single being, how was the division into the two sexes effected? Could Natural Selection do this? for in order that the male and female should be fitted for their respective parts, a very elaborate adaptation of their bodily organs would become necessary; but they must be furnished with these at once in order to fulfil the purposes for which they were made, and nature could not afford to stand still for several generations while Natural Selection was perfecting them, even if it could possibly perfect them at all.

The instinct which by the process of sexual

connection sets in motion the whole machinery by which Life in its different forms is renewed and perpetuated, is among the strongest in the economy of nature. How it could spring from Natural Selection alone is beyond my comprehension; but it is also to be noted that it is not only implanted in the vast majority of living beings, comprising all those raised above the very lowest types, but that it is regulated by certain laws. This instinct operates through a desire attracting the two sexes to each other. How is it that this appetite is always confined within its proper limits, and is never found in animals (in a savage state at least) to lead them to deviate from them? If the sexual instinct were to be widely diffused, a sort of chaos in animal life would be created. How wonderful is it that the male and female of each species are only drawn toward each other! In a state of nature this law is absolute, but in animals domesticated by man it would appear that it admits of some irregularity. Now here it would seem that Intelligence must be apparent: as I have before observed, foresight is a convincing proof of the presence of Intelligence. In the machinery for the production of Life through sexual connection, which is one of the most mysterious of all nature's secrets, two laws are traceable: the one, general, almost universal, that which restricts the sexual instinct to the male and female of each particular species. In a state of nature this law is absolute; in a state of domestication some irregularity appears, chiefly caused by the subagency of man interrupting to a certain extent the course of nature. But here another law steps in protecting the different races of living beings from that confusion which would result if such power had been left unrestricted. Some of the domestic animals, of different though very similar organizations, are found, under the direction of man, to be capable of sexual intercourse and of bringing forth an offspring; but here steps in another law of nature, a sort of second safeguard or barrier erected against the indefinite multitude of species: the issue of these connections is always barren. Does not this argue foresight in the Great Law Giver? How could Natural Selection alone create such prospective limitation? But if a future defect in the machinery for the renewal of Life is thus foreseen and guarded against, then there must have been a Being able to foresee and competent to provide the remedy. Natural Selection cannot foresee; the very hypothesis itself denies it such a power.

How different do these philosophers argue in the case of Providence and of Man. The attention of Geologists has of late been directed to the discovery of a large number of stone spear-heads at a very great depth in the marshes near Amiens. These have occasioned quite a revolution in the previously received Chronology as to the date of Man's birth. It is calculated that, at the depth at which they were found, some thousands of years must have elapsed since they were placed there, and it is at once pronounced that these rough spear-heads were fashioned by the hand of Man. I have seen a considerable number of them, and I think that they bear out this opinion. They are uniformly made, of exactly the same size and pattern, and very well adapted for the purposes of that description of weapon. If they were copied in iron and fixed to a staff they would be formidable in hand-to-hand encounters; they appear to have been chiselled out of flint or some hard dark stone with considerable neatness of execution, and bear the marks of the chisel over their whole surface. Now these rude weapons were not found in proximity to any relics of man;

there was nothing beyond their own construction to show that they were his work. They were immeasurably inferior in beauty of design or workmanship to a honeycomb or a thousand other natural productions, and yet Geologists at once came to the conclusion that they were of human production. There is a character of design stamped upon them, and the workmanship is of such a nature as at once to lead us irresistibly to such a belief; and, yet while we are surrounded by works exhibiting in a thousand times a greater degree proofs of wisdom and power in their formation, we refuse to acknowledge the hand of a Higher Power, and take refuge in this incomprehensible theory of Natural Selection.

The stamp of intelligence and design appears to me to be as unmistakably imprinted in a thousand instances upon the whole economy of animal and vegetable Life as it could have been upon the stone spear-heads; intelligence is recognized in its effects. I cannot conceive it possible that works bearing the indelible character of design can possibly have been created by the mere fortuitous operation of unconscious or material influences. Mr. Darwin should prefix to his volumes as a motto the parody in the "Rejected Addresses" from Dr. Busby's "Lucretius:"

> "I sing how casual bricks in airy climb
> Encounter casual horse-hair, casual lime."

I cannot conceive that any lapse of time, without any Intelligent guidance, could possibly have produced all these fruits of the highest combination of wisdom and power which constitute the phenomena of Life. I should as soon imagine the possibility of a ship building itself, rigging itself, and navigating itself from the Thames to Melbourne, without shipwrights, captain, or crew.

We must also remember that Life in its various forms is not the whole, but only a part of the works of Creation. These are certainly not all to be accounted for by Natural Selection; but if Intelligence is to be traced in any one of them, then the existence of Intelligence as a governing and creating power is established, and it is almost self-evident that the mighty power which is proved in any one operation of Creation must pervade the whole. It is not Natural Selection which guides the Electric fluid—it is not Natural Selection which dispenses Light and Heat—it is not Natural Selection which has created and upholds the

Solar System. If the existence of the Great First Cause be shown anywhere else, we may surely credit it with having also originated the phenomena of Life.

I know not how far Mr. Darwin may carry his views, or by what process of reasoning he may reconcile his own theories with any recognition of a Divine authorship; he seems faintly and obscurely to intimate the possibility of such an agreement, and he quotes his correspondent before referred to as entertaining such an opinion. It appears to me, however, that the very essence of his doctrine consists in its sufficiency to account for the existence of all the diversified forms of Life by the independent operation of secondary causes. I have no wish to invoke in opposition to him any spirit of intolerance or hostility. I have remarked, in a previous passage, that true Science is irresistible, and that it must be accepted, with all its consequences, wherever fully established. But where these consequences are so momentous (not to say awful) to all those principles upon which mankind have reposed, we must be permitted to scrutinize them in no very friendly spirit. There is no use in disguising the matter, or in shrinking from those conclusions which unavoidably follow from Mr. Darwin's writings. They are totally subversive of religion—a pious Darwinian would be a contradiction in terms—all those hopes which are identified with that sacred sentiment, all those restraints which our feeling of accountability to Heaven enjoins, would be swept away: a moral revolution would be effected, and that revolution would be subversive, not constructive.

Has Mr. Darwin succeeded in establishing his theory upon a basis of proof which must convince the understanding? To me it does not appear that he has succeeded in forming even a plausible one. In order to carry out his system, and to give time for the operation of those agents of Natural Selection and Sexual Selection, he finds it necessary to press into the service the whole series of Geological periods about which we know so little; at the same time he constantly admits what he calls the "Imperfection of the Geological Record," as an apology for the inconclusiveness of his arguments. Our existing world, with its short experience of some 4,000 years, affords him no foundation. Here we everywhere see what appears to our benighted reason irrefragable proofs of design; here we everywhere see forms of Life the most varied yet everywhere the most adapted to their several purposes; but nowhere do we see those agents of Natural Selection or Sexual Selection effecting any considerable results. The Earth does not exhibit at this epoch any instance of a new creation. Varieties may be produced by the skill of Man, which is itself a sort of Intellectual First Cause, but a new creation of the animal kingdom can nowhere be produced. All those products of the most wonderful and elaborate contrivance must have been gradually evolved out of nothing, by a process of Natural Selection which Mr. Darwin says may have taken millions of years to accomplish, and 120 generations is all that we can possibly look back upon. How does Mr. Darwin prove the connection between all these periods?—upon the sole evidence of fossil remains which come down to us, by his own showing, in a broken and disjointed condition. When we confine our observations to the existing Geological period, where can we trace the action of these agents? Mr. Darwin dwells upon the variation effected by breeders in domestic animals; all those varieties (which exist after all to a very small extent) are only found among animals of the same kind, and which breed and multiply with most perfect facility. Wherever a cross takes place between animals of a different although very similar species, as in the case of the horse and the ass, it is always a result of the contrivances of man, and never occurs in wild animals. Many varieties of the cat kind appear more nearly allied to each other than is the horse to the ass, but sexual intercourse never takes place between them. A tiger and a lioness differ very little—perhaps in the case of the lion he might alarm the tigress by his magnificent mane — but the lioness, who has no such distinguished appendage, does not differ very much from the tigress except in the color of the hide, far less certainly than the horse differs from the ass; yet neither between these animals nor any other species, in a wild state, does any sexual intercourse ever take place, or is any offspring ever born. So far as our experience extends, the barriers between the races are never passed.

We must persist, although Mr. Darwin objects to it, in pursuing an analogy between the works of Man and those of the Deity. Immeasurably inferior as are the productions of human invention, yet we know beyond dispute that these are the productions of Intelligence. If the word "Intelligence" has a meaning at all, a ship or a railway engine, or a watch or an Electric Telegraph, are productions of Intelligence; yet although the highest efforts of the human mind are immeasurably inferior to the works of Providence, still are they not the same in nature, and do they not only differ in degree? Can we say that a telescope is the work of design, while the eye has been evolved by Natural Selection and produced without an Intelligent Cause? Is there nothing to be accorded to that universal instinct of humanity which turns to a Higher and unseen Power? Is it not much easier, much more credible, to believe in an Intelligent First Cause as the Great and Invisible Power directing all things, than to construct by so elaborate a process works bearing the stamp of the Highest Intelligence and Power out of mere brute matter? In his recent work on the "Descent of Man," he applies the same process of evolution by Natural Selection and Sexual Selection to Man, and traces his existence in his present state to some species of ape or monkey from whom he has originated. I do not know if we have any ground for taxing Mr. Darwin with inconsistency in pushing his system to this length: it seems quite as inconceivable that the ape should have originated in a mass of lichen on a rock, as that Man should have sprung from the ape. The fallacy appears to me to exist in the assumption of a power of self-creation, or of self-development anywhere, which we are utterly unable to discover in nature. If we ask, what is Life itself? we shall find that it is a question quite beyond the faculties of man to answer; Life, like spirit, like matter, like the nature of the Deity, like the difference between reason and instinct, like the nature of the union of the soul and the body, is among those primary ideas which it is only given to Man to know and recognize by their effects, but the essence of which it is quite beyond the power of his faculties to conceive. We know that a stone does not possess life, and we know that the moss which covers it does; but we cannot explain in what lies the difference: all that we know is, that this difference could not have been imparted to the moss by itself—some external influence must have created it. By no conceivable process of reason or imagination can we arrive at the conclusion that Life is inherent in some bodies and not in others, or that it can exist without having been created. But if we come to this conclusion, it appears to me it disproves all Mr. Darwin's theory, because if we admit Creation as a necessary first step, and that such Creation is the act of a Power separate from or superior to the thing created, then it would follow that the thing created would be derived in all its subsequent stages from that Power. It is as great an act of Omnipotence to create Life as to form and fashion it after creation.

Mr. Darwin's first step in his theory of Natural Selection is what he terms the Struggle for Existence arising from the ever-pressing necessity for food acting upon all animated nature. We grant to him the truth of this first proposition. Life, as it exists upon this earth, is dependent upon air, water, and food; and as the supply of the two first requirements is practically unlimited, the amount of Life cannot exceed the quantity of the third. But Life is reproduced in much more rapid ratio than food can be augmented. If the countless forms under which it exists were suffered to increase, unrestricted by any check, they would soon overtake the supply of food, which they could not however pass, and the progress of reproduction would thereby be violently and suddenly arrested. But Nature interposes various checks regulating and controlling the operation of this principle, and preserving its system from the rude shock of such a collision. One great means for effecting this is, that so large a number of the animals and insects prey upon each other, and any excess in one portion of the Zoology of the Globe is immediately arrested by the action of those others to which it supplies the necessary aliment. Diseases, severities of climate, and other causes, contribute to diminish this redundancy wherever it exists. The necessity for food may be compared to the mainspring of a watch or other complicated machine: it sets the whole in motion, and regulates the movement, which would otherwise fall into confusion. We see its influence most distinctly in the case of man, whom it obliges to labor: it is at the root of all the movements and operations of that complicated machine which we call Civilization. It is not, therefore, upon this first term in his proposition that I am disposed to take issue with Mr. Darwin. He reproduces the theory of Mr. Malthus, which has always appeared to me quite incontrovertible, and through the agency of which so much of the economy of Life appears to be carried on. Mr. Darwin affords us some strong instances of the working of this principle. He states that the elephant lives for ninety years, and during that long period only gives birth to young three times, producing a pair at each birth. This animal is supposed to be the slowest of all breeders, and yet Mr. Darwin states that at this rate there would be fifteen millions of elephants at the end of the fifth century. If such be the wonderful increase of this slow-breeding elephant, what would be that of the herring, producing millions of spawn every year? We do not therefore dispute Mr. Darwin's first proposition, that all Life tends to a far more rapid increase than the resources of food would be long adequate to support, and that this struggle to maintain it has a great motive power in the World. It is the second step in Mr. Darwin's theory, the proof of which it seems to me he fails to establish—that this struggle for food is the cause of all those countless varieties of organized beings which people the earth. We do not dispute that the progeny of a pair of elephants in five hundred years might amount to fifteen millions, or that the spawn of a single herring might in the same period produce a number of descendants which it would tax the ocean to contain or the powers of arithmetic to enumerate. What we do dispute is, that by any chain of descent, of which we have knowledge, the elephants can have been the progenitors of the herrings, or the herrings of the elephants. This I conceive to be Mr. Darwin's theory, for although he may not undertake to affirm in what degree of consanguinity the herrings and the elephants stand to each other, yet he does assert that all organized and animated Life flows from one source, and is, in fact, related together. He also conceives that this relation follows from no action of a Superior Will, but from an involuntary and unavoidable sequence of cause and effect. Now the first part is proved with figures; it is arithmetically true, indeed, and proves itself; but his second proposition does not follow the least from the first; it is a question of fact, and are his conclusions supported by the evidence of facts? It is here that, it appears to me, Mr. Darwin's long chain of hypothesis and supposition commences which he substitutes for facts. I revert to a proposition which I have laid down previously, that we can only proceed by the light of our own faculties and powers, and by the knowledge that we possess; we cannot step beyond that circle. Now all the innumerable forms of Life, all the beings adapted to their several conditions, all the means and appliances by which they breathe, move, and have their being, appear to me to be stamped with the unmistakable character of design. There must be the mind to conceive and the power to execute, in order to produce the results before us. I do not see how it is possible to deny this; if we examine any one of the works of Creation, whether animate or inanimate, and if we analyze the meaning of the word invention, I cannot understand how it can be denied that all those vast and multifarious arrangements of parts to a whole, and of means to an end, as exhibited in all the provinces, whether of animate or inanimate existence, can fail to bear out this truth. Mr. Darwin may say that all these appearances are illusory and deceptive, that the whole of that wonderful machinery by which living beings, from their lowest to their highest forms, have been produced in all their marvellous combinations follow from his motive power of Natural Selection and Sexual Selection alone. But it is for him to prove his system. The first and obvious character imprinted upon the whole is that of invention, emanating from a superior Power, a superior Intelligence, and a superior Will.

I have remarked in a previous part of this

Essay that all our progress in knowledge and in Science is attained by three methods: the first by the application of the exact sciences to the determination of all those questions to which they can be applied; the second, the evidence of fact; and the third, experiment employed to elicit fact. Now the first step in Mr. Darwin's theory may be regarded as founded upon the first of these methods; the ratio of increase in living beings is an arithmetical problem which follows from the premises. But his second proposition—that all the countless varieties of organized Life upon this globe are derived from the action of two principles, that of Natural Selection and of Sexual Selection, is a proposition which cannot be mathematically demonstrated, and which must be established, if established at all, by the evidence of facts. Do Mr. Darwin's facts prove his theory? Let us recollect for a moment the immense and complicated variety of the conditions of Life upon this Globe, and the wonderful adaptation of Life in its different forms to such varying conditions: for instance, aquatic and terrestrial Life, where the conditions are so dissimilar as to be absolutely incompatible with each other, and yet the organs of the living inhabitants are so constituted as to be exactly suited to each state. Let us look at the birds of the air; the genius of man is fertile in mechanical inventions, but he has never been able to construct a machine by which he could fly; yet the wings and feathers of birds are mechanical contrivances, which no human skill has ever enabled us to imitate. It would be superfluous to multiply examples of all the marvels of the animate and inanimate world which surround us; no one can better appreciate them than Mr. Darwin, whose peculiar studies have constantly led him to examine them. He must therefore have at any rate satisfied his own mind of the adequacy of his two principles of Natural Selection and Sexual Selection to account for all the phenomena which he of all men might be supposed least likely to undervalue.

Nevertheless, is his reasoning of the second step satisfactory? and does he establish the connection between his two principles and the facts which he ascribes to their operation? In the first place, his actual experience must be a very confined one in comparison with the extent of the field which his theory embraces; it comprehends all time, but the living world alone and the scattered fossil remains of extinct geological periods can be brought within the sphere of his actual observation. But if I understand his work, the living world alone only exhibits the ultimate and latest effects of these principles; they have worked in order to produce them during countless ages, quite beyond his view. All the confirmation which actual observation can afford must be sought in these latest and most recent periods, whereas the working of his system has been carried on in the most remote. But he does not appear to cite instances of the results of Natural Selection as falling within later experience; indeed, he seems to require such vast periods of time for their operation that they quite pass the bounds of any human observation; we cannot see more than the very faintest indication; while, on the other hand, we everywhere encounter what appear to be fixed laws of nature, which are opposed to them. To begin with his first motive power — the Struggle for Existence — there seems to be no evidence to show that this struggle for existence leads to any transmutation or fundamental alteration of species; the primary effect of an insufficiency of food is to cause the diminution or extinction of the race, but it is not obvious how it can supply the place of invention, and transform the animal into one better adapted to new conditions. All the existing world of animated nature seems to be divided by hard and fast lines into different groups, which admit of no amalgamation; everywhere carnivorous and herbaceous animals—birds and quadrupeds, inhabitants of the land and of the ocean, all the countless varieties, races, and species—seem to be divided by impassable lines of demarcation. Mr. Darwin himself quotes numerous instances; he speaks of the humble-bee and other bees so nearly allied in structure and yet never mingling together or losing their distinctive characteristics. The great law of sexual connection seems to be that like seeks like, and the whole of Mr. Darwin's theory of Sexual Selection seems to be founded upon the hypothesis that this law is frequently broken or transgressed. But we cannot find any well recorded or defined instance in which this law has been violated, and a new race sprung up in consequence. It does not appear, in fact, that he ever traces any of these new formations to their commencement. The origin of all those different races, species, and varieties in the zoology of the earth seems to be quite beyond the reach of our mental vision; experience cannot be said to have brought any confirmation of Mr. Darwin's theory, and most of the known laws of nature contradict it. There is no proof that the struggle for existence has ever occasioned or does tend to create any modification in existing species, or that there are any irregularities or deviations from the laws of sexual intercourse sufficient to account for the vast variety of the races and species on the globe. The natural history of the earth worked out upon Mr. Darwin's two principles, and excluding the direction of Intelligent Design, is to me perfectly incomprehensible and incredible. Now, admit the agency of an overruling and Intelligent First Cause, and all appears to be simple and easy.

Mr. Darwin objects to any analogy drawn from the powers or faculties of Man; but I contend that it is only from this comparison that we can gain any conception of those higher powers which regulate the Universe. Let us contemplate some of the mighty efforts of his faculties recently displayed—the Electric Telegraph, for instance. By this invention man can communicate his thoughts with the speed of light to the most distant quarters of the globe; but it is not the electric fluid passing along the wire, nor the signs denoting the letters of the alphabet, nor the letters of the alphabet, which of themselves express no ideas, but it is the Intelligence of Man, combining all these different agents for a purpose, which Intelligence alone could plan, and an Intelligent recipient could alone benefit by. Although we cannot trace the subtile agency by which the government of the world is carried on, yet it is quite within our power to conceive; just as the intellect of man and his power over the material world effects such wonders as the Electric Telegraph, so a Higher Power and a Brighter Intelligence may effect all those operations of nature which we see around us. There is nothing very difficult to understand in its operation; we cannot trace its subtile agency, but there is nothing at all difficult to reconcile to our reason the supposition that such agencies exist. If we grant the existence of the overruling Power directed by Intelligence, the whole scheme of Nature, and the forces which move, guide, and control it, are perfectly conceivable. But this theory of accounting for every thing by the operation of secondary causes, by whatever name they may be distinguished, whether Natural Selection, Sexual Selection, or any other, is to me perfectly incomprehensible. I can neither trace their operation in the actual world, nor can I imagine it by any exercise either of reason or of fancy.

There is one consideration with reference to this subject which appears to me entitled to considerable weight. It is this: if the operation of the Divine Will is supposed (as Mr. Darwin in parts of his work appears to suppose It) to have any influence in the economy of nature, It must have the whole power; we cannot conceive It to be partial, indirect, or divided: if he admits It in any degree, he must allow It to be universal, direct, and supreme. A partnership in the exercise of Omnipotence is the most irrational of all suppositions.

In conclusion, it appears to my comprehension that Mr. Darwin's theory for accounting for all the innumerable varieties of animal and vegetable Life in this globe is not proved. It must depend upon the evidence of facts; and facts are so far from bearing it out, that they seem in many necessary particulars to be altogether wanting, and in others to contradict his premises. They are quite insufficient and inadequate to accomplish the results ascribed to them.

First, he assumes that these various forms were transmuted or evolved one from the other in an endless succession; that each variety springs somehow from another variety; in animal life particularly that each variety of animal form is produced or evolved from some other different animal form, and that these new varieties have arisen from some changes in cognate or nearly similar species. For example, he supposes that Man has sprung from some variety produced either by Natural or Sexual Selection in those apes that physically bear the most resemblance to the human frame. But this theory is totally unsupported by any positive fact. No naturalist ever knew of any such variety thus produced, nor of any variety whatever produced by any analogous process in any other animals. Mr. Darwin seems to consider that he has proved his case when he has shown that all animal and indeed vegetable Life is divided into classes, genera, and species which do more or less resemble each other. Such would in all probability be the case, if every one were a distinct creation. Creation would probably be carried on in groups which, though independent of each other, might be classified under distinct heads. But Mr. Darwin is totally unable to supply the next link in his chain, by showing that in any one case these species have become intermingled. The laws of nature appear to be quite fixed and absolute in this respect, and no change can ever be traced by which any one species is merged in another. In order to prove this theory, Mr. Darwin should have been able to cite cases in which new forms of life have been thus evolved; but he fails to do so, and cannot cite a single case confirmatory of the process by which he supposes the whole zoology of the globe to have been metamorphosed. In the next place, were we for the sake of argument to concede that this process of Evolution by Natural Selection and Sexual Selection had produced all the innumerable varieties of life which exist, they would bear the unmistakable evidence of design, and therefore that they would be the manifestations of a Great First Cause. These secondary agencies, to which he refers, would in that case be the mere instruments of His Will. We might as well suppose that the Transfiguration by Raphael or the Communion of St. Jerome by Domenichino were the works of the paint and the brushes which delineated them, or even of the hand which used these instruments, and not of the minds of Raphael and Domenichino.

The works of Nature all imply design; and the secondary agencies evoked by Mr. Darwin can have no design, and must therefore be subordinate to some Higher Power that has. The belief in the agency of a Great Intelligent Omnipotent First Cause fully satisfies the understanding, and alone enables the mind to repose in the certainty of conviction. Faith in the existence of a direct though invisible communication is an essential part of our moral nature; it is interwoven with all our best hopes and highest aspirations, and is necessary at once to happiness and to virtue.

To sum up my objections to Mr. Darwin's theory: 1. He either does or he does not refer to the agency of an Intelligent and Omnipotent First Cause. If he does so assign it, then all the complications of his machinery, his Natural Selection, his Sexual Selection and Evolution, by which he accounts for all the endless variety in the forms of organic Life, are only so many additional difficulties in the work accomplished by the Deity. It becomes the more necessary than ever to believe in design; because a governing power directing the whole becomes more imperative in pro portion to its complication. If he excludes all reference to the Intelligent First Cause, then this chain of totally distinct organizations issuing one out of another becomes quite incomprehensible. Unless we suppose design and a presiding direction, nothing but chaos could result.

2. I do not understand how Mr. Darwin can support his system by the evidence of fact. He supposes that in an incalculable number of ages one organism gradually melts into another, but he nowhere can point out any one case in which such a transformation has taken place. Slight variations in some of the domestic animals, which he adduces as proofs, seem wholly insufficient, since they are only to be found among those slight varieties who freely cohabit and breed together. Where these slight varieties assume a more fixed character it is always in obedience to the subordinate but still Intelligent Will of Man. Mr. Darwin's whole arguments seem to be confined to showing that different species resemble each other very nearly and are separated only by apparently slight lines of demarcation; and then he assumes that these lines are in the course of ages overstepped and new species thus fashioned, but he never does show that these slight lines are so overstepped. With all his wonderful knowledge of Natural His-

tory, he cannot prove this from fact. On the contrary, his work teems with instances in which the very slightest different characteristics are always permanent, and no intermingling takes place. For example, in two or three different species of bees, so resembling each other in every particular, nourished by the same food, living in the same fields, constructing almost similar hives, and yet separated from each other by a few marked differences, and remaining always distinct. This, I think, is a fatal objection to Mr. Darwin's theory, that he always supposes that at some remote period such causes have separated different organisms, but he never adduces a fact to support this theory, and all the known laws of nature seem to contradict it.

Another modern author has bequeathed to English literature an interesting work on the "History of Civilization." It must be a matter of regret to the world of letters that Mr. Buckle was so soon removed from among us in the midst of those labors which bear the stamp of so much talent and originality of thought. But while all must admire his early promise, and regret his loss, such a tribute does not involve any assent to his opinions, or conversion to the doctrines which he so boldly enunciated. Mr. Buckle's genius was essentially speculative; he delighted in inventing new theories, and was more ambitious of originality than careful of soundness. He did not, like Mr. Darwin, push his inquiries beyond the very existence of Man; he was satisfied with carrying them into the darkness of the prehistoric times. Still to a certain degree these two philosophers treated of the same subjects; and I cannot leave Mr. Darwin without a few remarks on Mr. Buckle's system.

In one material respect they contradict each other, and Mr. Darwin's positions and Mr. Darwin's conclusions are totally subversive of one of Mr. Buckle's main principles. The latter declares, without stopping altogether to prove it, that the primitive state of Man was one of absolute equality; that every member of the human family came into the world the equal of every other; and that all the differences, whether of nation or of race, are the results of external circumstances influencing their subsequent condition. Now Mr. Darwin maintains a totally different doctrine. In his chapter on the "Races of Man" he distinctly states that they present varieties so strongly marked, as to have almost justified some naturalists in regarding them as different species; and he has applied to them the term "sub-species," indicating that although the distinctions are not so broad and absolute as to justify their being ranked as distinct species, yet that they are too widely apart to be classed merely as varieties, and he has therefore invented a sort of middle term of "sub-species" by which to characterize them.

Now this is totally irreconcilable not only with Mr. Buckle's premises, but with all his subsequent deductions, which proceed to explain by some very ingenious but far-fetched hypotheses the causes of these differences which, according to him, were not innate, but have been subsequently grafted upon Man's original nature. Mr. Buckle lays down that all the diversities found in Mankind spring from differences in "Climate, Food, and Soil, and in the impressions made upon the mind by the different Aspects of Nature." With regard to the first three, there can be no doubt that they have a considerable influence upon the character and condition of nations; but Mr. Buckle is inclined, as most inventors of systems are, to bring every thing within their operation. If Mr. Darwin is correct in supposing that there are natural differences in different races of Man, which are either coeval with their birth or spring from some very remote and antecedent state, then these natural differences would inevitably act upon the future progress of the different races, and would be a cause of difference quite independent of either of the four to which Mr. Buckle would alone confine them. Mr. Buckle is very fond of laying down some absolute rule to which he conceives the whole history of Man conforms. As an instance, we will take his position—the fertility of the soil in an early state of society always leads to a small wealthy class who are absolutely despotic, and who crush and oppress the bulk of the population. He instances Egypt, but Egypt is not singular in being despotic, or in sharing that principle of government with fertile countries alone. The whole of Asia, from the earliest records, seems to have been governed by despotic rulers, yet many parts of Asia are rigorous in climate and sterile in soil—a proof that there is no exclusive connection between fertility and despotism, since despotism is quite as absolute and quite as cruel in the less-favored regions of the north. Were Zengis Khan, Tamerlane, or Attila, one whit less arbitrary and tyrannical than the Pharaohs? Were the ancient Czars of Russia very humane and liberal rulers? As I have noticed elsewhere, quoting from Montesquieu, all Asia has been immemorially governed by despotisms, and the principle of fear has been the sole and universal engine by which authority has been maintained.

Some of the most fertile regions in the World are the banks of the Mississippi and the great rivers of the North-American Continent. Mr. Buckle broaches a theory, that to produce fertility it is necessary that heat and humidity should coexist in the same country, that heat is confined to the Western coasts of America beyond the Rocky Mountains, that humidity, without heat, is found in the Eastern portion of that Continent. Nothing can be more erroneous. It is possible that there may be greater aridity in some parts beyond the Rocky Mountains, but so far from there being any insufficiency of heat in all the Southern Provinces of the North-American Continent, the Floridas, Louisiana, Alabama, Texas, Georgia, and Virginia, are very hot countries, and there is quite sufficient sun, combined with their moisture, to produce the most luxuriant vegetation. The banks of the Mississippi are quite as favorable to human life, and quite as capable of supporting an enormous population, as the lofty table-lands of Mexico; and other causes must be found than those alleged by Mr. Buckle why they did not become the seat of a population as numerous and as civilized as either the Mexicans or Peruvians.

Mr. Buckle's theories respecting the Aspects of Nature are still more unsound and visionary. He supposes that the terror inspired by the repulsive aspect of tropical countries, and by the alarming convulsions by which nature is visited, press upon the mind and imagination of Man, and induce him to be weakly, superstitious, and selfish, while more engaging and cheerful aspects brighten his faculties and elevate his moral nature. He compares Greece with India, but Greece is by no means so deficient in the terrible: her mountains are high, often rugged, and in earlier times her forests must have frowned over the blue seas of the Mediterranean. India is indeed a tropical country, but it by no means presents those terrible features with which Mr. Buckle invests it; there is no part of the world in which there are fewer earthquakes, its soil is not volcanic, and the general character of the Peninsula is more beautiful than savage; the lower valleys of the Himalayas are among the loveliest spots on the globe; and there is no conceivable ground to a man with ordinary observation for supposing that Vishnu and Siva should have been conceived by the imagination of Man, contemplating these beautiful and picturesque localities, while Parnassus should have given birth to the "Iliad."

Mr. Buckle appears to have had a great facility of adapting his theories to circumstances. In the Old World a fertile soil leads to redundant population, cruelty and oppression in the upper orders, and slavery among the mass of the people; but in the Eastern portion of the South-American Continent vegetation becomes too luxuriant, it overpowers man, and stifles him beneath the very redundance of its productions. He is engaged in a war with Nature, in which he is crushed and subdued by the very vigor of her productive powers. All this, however, is a mere dream; Brazil is one of the most flourishing States of the New World, and its population is rapidly increasing.

There is one feature common to the writings both of Mr. Darwin and Mr. Buckle which is to be regretted—they both of them seem to ignore, if they do not altogether deny, the existence of a First Cause. Secondary causes are always with them the only springs of motion. This gives a repulsive chillness to all their theories, which the mind instinctively rejects.

PROGRESS AND CIVILIZATION.

There is no one idea which so universally pervades the public mind of Europe as that of Progress. The Progress of the Nation, the Progress of the Age, the Progress of Civilization, the Progress of Society, the Progress of Mankind, are phrases which meet us everywhere. We find them in the columns of every newspaper, in the articles of every Review, in the philosophic pages of Buckle and Macaulay, in Social Science Congresses, in private circles, in Parliamentary Debates, and even in the Pulpit. Nor is there any thing the least surprising in the general prevalence of such ideas. We whose lot has been cast in this nineteenth century have had the high privilege of living during a most eventful epoch in the history of civilized man. We are often referred to the fifteenth and sixteenth centuries as the period when Europe was aroused from the lethargy of the Middle Ages, and when various discoveries and inventions imparted a new impulse to civilization. We read of the inventions of printing and of gunpowder, of the discovery of the properties of the Magnetic needle, and the consequent improvement of Navigation. The progress thus made in this art led to the discovery of the New World and of the passage by the Cape of Good Hope, while from the great astronomical discoveries of Copernicus and Galileo in the same period may be dated the birth of modern Science. There cannot be a doubt either that these great events did give an extraordinary stimulus and momentum to the advancing movement of that age, or that these powerful agents are still at work after the lapse of between three and four centuries. They have lost none of their original force, they are not spent or exhausted; but, on the contrary, are going through a process of development, which is adding to their strength, and extending the sphere of their operation. We are now better able to appreciate their vast importance, and to measure the immense results which have flowed from them, than when they originally burst upon the World. We have inherited all these means and resources in their full force and activity; but we have added to them new ones, not less potent, not less wonderful, not less signal triumphs of the intellect and genius of Man.

In the foremost rank of these new achievements of the inventive genius of Man, the first place will be assigned to the application of the motive power of Steam to every description of manufacturing industry, and to the three great purposes of river and ocean navigation and railway communication, and immediately after these follows the still newer invention of the Electric Telegraph. We advance so rapidly, and these great inventions of our age have now become so thoroughly established as part of our daily existence, that we recall with a sort of surprise that they were all unknown during the first twenty years of the present century. Those still only on the threshold of old age can perfectly remember the days when Sailing-packets, Mail-coaches, and Post-horses were our means of locomotion. Steamboats were not established between England and France or Ireland until after the year 1820, and the first great Railway line for Passenger-traffic, the Manchester and Liverpool, was opened in 1830, acquiring a melancholy celebrity by the tragical death of Mr. Huskisson. The new invention did not make any rapid progress at first, and it was not till the ten years from 1840 to 1850 that our principal lines were completed. Since that time these mighty works of human skill and power have been extended with wonderful rapidity. Railways traverse, more or less completely, every country in Europe. The great Continent of North America is intersected by them. In India they are in course of construction. We may safely assume, however the process may be delayed by the immense amount of Capital they require in their formation, that another half century will see them everywhere coextensive with European civilization, and bringing new regions and semi-barbarous peoples under their beneficial and humanizing influences. The progress of Steam Navigation, as it was earlier in date, and is created with less amount of Capital, has been still more generally diffused. Ships navigated by Steam, either wholly or in combination with sails, traverse every sea and ascend every river. They are to be found in the Chinese waters, the Eastern Archipelago, the Caspian Sea, and the River Volga.

No one whose attention is alive to the

present temper of the times can fail to remark that this idea of Progress has acquired a predominance over the public mind which it never before exercised. It has become in itself a moving force in human affairs. I mean that the universal impression that great changes are taking place, and that Progress is a law of Nature, has in itself an active influence. There is, however, something vague and indefinite in the current impressions which are so widely scattered throughout all classes of the community. Men echo the phrases, "The Progress of Society," "The Advance of Civilization," "This is an Age of Progress." They use these general expressions without any very distinct or accurate sense of their purport, or any attempt to explore the depths of the vast subject comprehended in these expressions. A civilized State is generally understood to describe a society which has attained a certain point of refinement and cultivation, in which the Arts and Sciences are studied and understood, where a regular Government is established, and where a system of laws administered with tolerable justice, and securing the fundamental rights of property, is established.

A rough, but not very incorrect classification is adopted in familiar conversation. The European Nations, and their descendants established in America and Australia, are alone admitted to the rank of civilized nations; the Asiatic Races, partly extending over the North of Africa, are semi-barbarous, or half-civilized, and the rest of the World is regarded as more or less in a savage state. These popular ideas seem of a superficial character. They do not answer the questions: What is Civilization? What is Progress? When and where did it commence? What are its causes, its laws, its duration, and its limits? No subject of human inquiry can be more extensive, more important, more profound, or better worthy of Philosophical Analysis.

What is Civilization? When, and where did it originate? These are questions which the popular and current acceptation does not satisfy. Civilization is used in a comparative sense. We are civilized in relation to some other States which are half-civilized, or not civilized at all. We are civilized now, in contrast with some former period, when we were ourselves uncivilized, and out of which we have emerged. What date shall we assign to that epoch? Shall it be the revival of learning, and the birth of Science at the close of the Middle Ages? Shall it be the Christian era? Shall it be the foundation of the Roman Empire, or the Babylonian Monarchy? All these are no doubt important stages in the History of Man; but when we have reached the remotest, we find that we have not yet attained the Zero of Civilization. Let us adopt another test, and try to discover that period in which what we at present call the arts of civilized life had their origin, antecedent to which Man was utterly rude and uncultivated, and from which we have derived nothing, and nothing has descended to us. But such an inquiry will lead us into still darker obscurity. We shall find that not only in Science, but that in all our Arts, in all the contrivances of human ingenuity with which our modern life is surrounded, whether the most familiar or the most elaborate, what is recent is constantly interwoven with the remote and unknown Past. All that we can trace is one continued and connected movement, the beginning of which extends far beyond the limits of History, or of Tradition, or of any known Record. We can only arrive at an approximate guess by examining ourselves, and by endeavoring to discover in what faculties of our nature this great distinguishing characteristic of Man is situated.

In illustration of this view, let us take one of the latest, and one of the noblest creations of an advanced civilization; let us select the "Warrior" man-of-war. How much of human skill, contrivance, invention, and Science, have been combined to produce that work; how many hands have labored, how many brains have toiled, how much of Art and Science, of mechanical ingenuity and skilled labor, have been put in requisition to launch that mighty construction upon the ocean? A legion of miners, iron-founders, ship-builders, carpenters, sail-makers, cordwainers, and artificers of almost every description, have united to frame her. Astronomy has given us the means of ascertaining her place in the trackless ocean. The great discovery of the fifteenth century—the Mariner's Compass—has enabled us to direct her course; the foremost invention of our own times—the motive power of Steam—has empowered us to impart resistless velocity to the mighty mass. Armstrong guns, conical shot, all the most recent improvements in Artillery, have been enlisted to augment her force as an Engine of War. Nothing more modern, no better type of Progress, could be chosen. But among all these modern inventions, among all these achievements of the fifteenth and succeeding centuries down to our own nineteenth, do we owe nothing to a much earlier epoch, to a date altogether too remote for us to trace its origin?

It would be useless for the Compass to point the right direction, or for Steam to propel her with untiring speed, if the rudder did not enable the steersman to govern her every movement, and to render all this mighty machine subordinate to his will. The rudder is a tolerably simple contrivance, but it is not altogether an obvious one. Rudders do not come by nature, or grow spontaneously out of the sterns of ships. The first rudder was probably hailed as a great invention, and might have immortalized its author, had there been any means of chronicling inventions. When did he live? What nation did he belong to? How long have vessels been steered by rudders? None of these questions can we answer. All that we know is that the rudder has descended to us from remote ages, and that it is now as necessary to the "Warrior" as either the Compass or the Steam-Engine.

If we push our inquiries farther we shall be able to discover the inventions of a yet higher antiquity, forming a necessary part of the basis on which this wonderful structure has been reared. Of what material are the sides, and the armament, of the "Warrior" composed? Of iron. And how is that iron ore extracted from the stone in which it is lodged? By the action of Fire. How is the Steam generated, which is her moving force? By the action of Fire upon Water. The element of Fire is one of the primitive constituent parts of the material world; but the power of kindling, managing, directing, and extinguishing fire, is an effort of human reason and invention. No animal we ever heard of ever lighted a fire or cooked its own dinner. The use of Fire was essential to the construction of the "Warrior," and when was Fire first employed by Man? We cannot answer that question.

In all that vast floating mass there is scarcely the smallest portion in the state Nature has produced. The raw material she has furnished has been so changed, moulded, and combined with other substances, by the various arts of Man's ingenuity, as scarcely to be recognized in its original condition. Human heads have planned, human hands have executed, all the parts of that great machine. The latest inventions of mechanical sciences—the Steam-Engine and the Screw-Propeller—contribute their aid to give her velocity. When she is launched, how much of Science is comprised in that art of Navigation, which has been slowly reaching its present advanced stage through centuries of Progress? The ground-work of Navigation is the Science of Astronomy; the spheroidal form of the Earth, its annual and diurnal motions, the declination of the Sun, the relative position of the Moon, the apparent angles with the fixed Stars, all are essential component parts of that Science of Navigation so necessary to the direction of the Vessel.

I have selected the "Warrior" as a convenient example of the truth of my assertion that Civilization cannot be dated from any known epoch, but that it is the gradual development of the faculties originally implanted in our nature by its Divine Author. There is an advantage in choosing this example, because a vessel like the "Warrior" stands singly. It is a creation in itself. The elements in its formation are thus more easily traced, and more distinctly marked than in more complicated results of the intellectual powers of Man. But every department of

Knowledge, or Science, every work of human hands, would afford, if subjected to scrutiny and analysis, similar results. In all it would be manifest that there is no branch of our acquirements or acquisitions which has not sprung from beginnings evidently stamped with the unmistakable character of human invention, but of which all record is lost.

Civilization, in its highest and most comprehensive sense, is the united accumulation of all Science, of all Art, of all the mechanical inventions, of all useful trades, of every part of knowledge which have been handed down to us. It is the sum total of that vast inheritance of acquirements of every description which has been bequeathed to us by all past generations of men.

This illustration, drawn from the "Warrior," leads us to the conclusion, that the whole history of Man forms one scheme, and that it is one of the designs of the Creator. Each part is linked to every other, and the whole forms one consistent and continued plan. Under this aspect, Man may be considered as a unity, as an individual, as a single chain, consisting of innumerable links. It is unfortunate that the present state of Geological Science does not afford us the means of arriving, even approximately, at the commencement of that period when Man first appeared upon the globe. The earlier Geologists seem to have fixed this at a comparatively recent period; but Sir Charles Lyell, and other later ones, have thrown great doubt upon their calculations, and assign a much remoter period to his origin. Whatever may be the date of it, we may assume that he was endowed then, as he is now, with the two gifts of Instinct and Reason; his Instinct appertaining to the merely animal part of his nature, his Reason constantly tending to elevate him above it. His instincts, we may conclude, were inseparably connected with his corporeal frame, or formed a part of him, and were identified with it and him, from his very birth. The higher faculty of Reason must have been developed far more gradually, and must have been some time in acquiring the government of his actions. The instincts of the lower animals serve for their preservation: were the instincts of Man at the very earliest commencement of his earthly course sufficient for his preservation? Did Reason, even from the beginning, lend its aid? or did the Power which created him aid the first steps of his infancy? Whichever may be the case, there must have been a period, and that a very brief period, after his birth, when his higher nature must have asserted itself, when his instincts became subordinate to his reason, and when the latter became the governing principle of his nature. Whenever that first step was made, the seed of Civilization was planted.

It is a distinguishing characteristic of Man that the child can receive new impressions from the parent; ideas not originally innate are formed in one generation; inventions are made, human ingenuity finds means to add to Man's command over some portion of the material world, and the power which Reason gives the parent of transmitting to his offspring any such new acquirements constitutes the first step in the Progress of Civilization. It is worthy of remark, that no animal possesses in the remotest degree this power of grafting by an intellectual process any improved methods of dealing with the material world which did not form a part of its original nature, and which were not prompted by an unvarying instinct.

There is a sect of philosophers always desirous of assimilating Man to the brute creation, and of identifying him with it by dwelling upon all those parts of his nature which he has in common with it. But this faculty of change, this gift of appropriating to himself new powers, and of extending his command over inanimate nature by fresh contrivances, and of subsequently transmitting them to his descendants, is possessed by him alone: no animal has the slightest shade of it, the most intellectual species of the brute creation can never make the slightest permanent advance, for they are entirely without this power of grafting a new habit upon their offspring. It is, I think, almost self-evident that the Progress of Civilization must entirely follow from this power of transmitting the acquirements of one generation to the succeeding one, which in turn passes it to another, enriched and augmented by its own. It must also be remarked that this process is a purely intellectual one, and is distinct from the transmission of qualities by hereditary succession. No one doubts that corporeal, and to a certain degree mental, qualities are transmitted by hereditary succession; every one is aware that family likenesses are long preserved, and peculiarities of constitution and even mental and moral characteristics are often reproduced by lineal descent. What is traceable in individuals is indisputable in nations and in races. It would be little short of a miracle if a blue-eyed, flaxen-haired, fresh-colored child were to be born on the coast of Guinea of negro parents: or if a little woolly-haired, thick-lipped, ebony-skinned negro were to be the offspring of a couple of Chinese parents at Pekin. Even in nations most nearly allied differences of race are distinguishable. Within our own Islands, English, Scotch, Irish, and Welsh, are marked by sufficiently obvious differences, both mental and physical. But all these differences spring from another source altogether; they are traceable to that mysterious process of nature, by which Life is renewed, through the intercourse of the sexes. But the process by which inventions, arts, and knowledge are transmitted from one generation to another is a purely intellectual one, and has nothing whatever to do with the laws of generation. There may no doubt be natures more or less adapted to receive these ideas, but the ideas themselves must pass from one individual to the other through a purely intellectual medium. Some children may learn to read more easily than others, but among the most cultivated nations children are never born able to read.

The principal means by which this advance must have been originally made, and the acquirements of the parents imparted to the children, must have been through the medium of the gift of Speech, another of the qualities possessed by Man alone. This power of communication between the different members of the human race is absolutely essential to Progress; without it the mind could never have expanded beyond the limits of the lower animals. Some small advance in mechanical skill or workmanship may be transmitted from an older to a younger generation by manual instruction: a carpenter may show his apprentice how to use a plane, a hammer, or a saw, without the use of words, or a sailor may teach a cabin-boy the art of knotting and splicing in perfect silence. But there is no real difference in principle. A certain mental process directs the saw or teaches a sailor to knot; it is equally an art transmitted by an operation of the understanding, though the means of communication are not quite the same. This mode, however, of transmitting arts from one generation to another is far more restricted, and is, indeed, of a very limited operation; it may be considered as merely supplementary and auxiliary to the mode of communication by Speech.

Language is at the root of all improvement; it is not only that we cannot speak without words, that we are altogether cut off from interchanging our ideas with each other, but without words we cannot think; let any one examine the operation of his own thoughts, and he will find that beyond the simplest ideas of form and color, sound or taste, which are the direct impressions made upon the senses, all ideas of a more complex character must be embodied either mentally or orally in words in order to enable the individual to conceive them. This consideration will assist us in estimating the immeasurable importance of that greatest of human inventions — the Alphabet. How curious that this great step in human acquirement is now synonymous with the very lowest rudiments of learning? The youngest child is expected to know its letters as soon as or often before it can speak, and yet it is the mightiest single stride that the human mind ever made in its progress toward a higher state of knowledge. Let us suppose the first inventor of the Alphabet, whoever he might have been, coming forward in some assembly of elders or sages, and exhibiting a handful of black marks traced upon any substance, and saying, "With these figures I can express all my wants, all my wishes, all my ideas; I can make them convey the most delicate shades of thought, and embody them in a durable shape; I can describe scenes and events with such accuracy that my auditors may almost fancy that they have themselves witnessed them; I can transmit to the most remote quarters of the globe and to the most distant periods of time narratives of the events which are passing around me." How difficult would the sages have found it to believe that so apparently simple a contrivance should be capable of such results. Probably no single individual ever did invent letters or establish such a claim to the gratitude of mankind. The process was probably a gradual one; hieroglyphics or cuneiform characters would probably among all nations be the more obvious method of recording events; a rude representation of the thing to be recorded, a battle or the erection of a building, would be roughly sculptured in some durable material; then perhaps some conventional signs might connect different events; but the idea of representing them by characters, not resembling or bearing any reference in form to the objects to be described, would be a much later process. Then would come the idea of connecting these characters with sounds, and gradually by successive stages arriving at a written language. But the superiority of this mode of expression would not at first be recognized, and probably it was long before hieroglyphics were insensibly and gradually supplanted by written words; it seems, indeed, that the two coexisted during the early ages of history; to this day the Chinese character appears to be a sort of combination of the two, yet it seems to be a form capable of expressing ideas with considerable facility; the Chinese are great readers, and their books are generally in the hands of the natives as they sit at their stalls and their shops. It is curious that not only are we ignorant of the name of any individual inventor of the Alphabet, but we do not know with any certainty among what nation the art of writing was first practised. This mighty invention, through which all the fruits of human thought, past and present, are transmitted to us; this invention by means of which we of these later times become acquainted with the Poems of Homer, the Histories of Thucydides, the Philosophy of Plato, and the Inspirations of Isaiah, seems to have stolen among the human race with so gradual and noiseless a step as to leave no trace of its first appearance.

Yet it has gradually developed into the great engine of human progress, connecting the whole human race together past and present, and conferring upon Man, considered collectively, a species of earthly immortality.

> "Non omnis moriar, multaque pars mei
> Vitabit Libitinam."

The thought so beautifully expressed by the greatest of elegiac poets, has acquired an expansion, possibly far beyond his own conception. He most likely limited his aspirations of future fame to the Roman Empire, which he might have conceived eternal, but the Roman Empire has passed away; yet, through the medium of printing, the Odes of Horace have obtained a wider celebrity than Rome could ever have given them, and will probably be the delight of the scholar, in all countries, for ages to come.

The invention of written language is in many respects a far more accurate and comprehensive medium of thought than speech itself. It is generally more the offspring of deliberation and reflection; it is more completely an intellectual process. Speech is often the almost unconscious utterance of some passing impression, but it is the mind and the will which dictate writing, and leave a permanent record of thought. It has exercised a wonderful influence in extending and perfecting the gift of speech itself; it has gradually imparted to it regularity, grammatical correctness, and copiousness. The perpetual recurrence to a fixed standard must have a constant influence in correcting errors, and giving to language greater precision as a vehicle of thought, while at the same time it must afford constant facilities for new modes of expression; by the perpetual extension it will be giving to the intellectual powers, it will be constantly stimulating the mind to satisfy its wants by coining new words and framing new forms of expression. It is impossible to estimate to what extent all the intellectual faculties of man have been enlarged by this great invention; but we cannot avoid perceiving that all the higher achievements of the human race are directly or indirectly attributable to it. It seems so necessary to us, so interwoven with our very being, that we can scarcely contemplate it in the light of an art or an invention; it seems to have become a part of our very nature. All accurate knowledge of mankind is coextensive with this art; every other means by which we try to read the Past is vague, speculative, and shadowy; they resemble at best the faint recollections of our earliest childhood, which we are scarcely able to distinguish from dreams. All those Sciences which treat of the structure of Man, or of his origin, or of the conditions through which he has passed, can only afford us external views of him. We study with more or less accuracy and success, just as we would study any other being distinct from ourselves, but written language enables us to identify ourselves with his inner nature, and to know him in some sort as we know ourselves.

Modern Science has been making efforts in different directions to penetrate through the mists which overshadow Man's origin; but it may be doubted if these inquiries have led to any certainty, or, even if these results were attained, would have any particular bearing upon our present condition, or our future prospects. Sir Charles Lyell has involved the chronology of primeval man in yet profounder darkness and obscurity; and the philological researches of Prof. Max Müller have given us but a shadowy glimpse of the races who are either altogether extinct, or who have become so blended together as to have lost all separate identity. But these inquiries must always possess a philosophic interest: the materials which enable us to study Man as he at present exists can only be furnished with an approach to accuracy since the dawn of the historical period. We may be told of the Aryans as our progenitors; but we know so little of them, that we derive little more than a name from these studies. But when we come down to the historical period, our knowledge of races and of nations acquires distinctness; we recognize them through successive periods, as retaining the same features, and it is wonderful how accurately their characteristics are reproduced during long ages.

The origin of Man is involved, and probably will always continue to remain, in impenetrable obscurity; but were even the most correct knowledge of it obtained, we should probably add little more to our stock of information. It would most likely be the history of savage and migratory tribes in the earliest stages of civilization. It might be compared to the biography of an individual in infancy, which would contain few or no indications of his future character. We possess perhaps all the materials necessary to the knowledge of Man from as early a period as there was any thing to be known; it is certain at least that the histories are abundant and increasing which may enable us to estimate his nature. One great subject of controversy meets us at the very threshold of our inquiries. It is the question, Whether the human race spring from one common origin, or whether the varieties of the great family of Man arise from distinct sources? On this subject physiologists are much divided. Some dwell upon the marked differences, others rest upon those features which are common to the whole race, and the dispute in the present state of our information is not easily settled. Without venturing to come to any decision upon it, we may safely assume that in either case the practical result at the present stage is the same; whether the various races who people the globe, the Europeans, the Mongol tribes, the Red Indians of North America, or the Negroes of Africa, sprung all originally from the same parents, and their differences arise from the operation of climate, modes of life, and other causes, through a succession of ages, or whether they were originally separate families, in either case the results at the present period are identical. These different races of men are as distinct from each other as different varieties of the same species among the lower animals. There is as much difference between an Englishman and a Negro, both physically and morally, as between a greyhound and a bull-dog. Their physical frames, their moral qualities, and their intellectnal powers differ widely, and the difference is not lessened whichever of the two theories is adopted. Nothing would seem to be more clearly demonstrated by the evidence of facts, than the existence of wide differences of character between races and between nations. It is just as certain and as indisputable to an unprejudiced mind as the diversity of individual character in private life. It would be as easy to contend that all men were equally tall or equally short, equally fair or equally dark, equally wise or equally silly, equally brave or equally timid, as that the characteristics of nation and of race were always the same. Of all the fallacies which have misled mankind in modern times, the greatest and most apparent, and perhaps the most mischievous, has been that of the natural inherent equality of mankind. The word Character, like most abstract terms, is not easy of definition, and yet is perfectly intelligible. We know very well what we mean when we speak of the character of an individual, the character of a nation, or the character of a race; but the exact signification of the word, which combines a certain union of mental, moral, and physical qualities, stamped with the impress of individuality, is not very easy to analyze with accuracy. We must assume, therefore, that they are intelligible when we say that, from the earliest ages of which we have any trustworthy record, distinct and marked features of character are traceable in the different nations of the globe.

One of the most remarkable of these is the Oriental or Asiatic type. This is the largest of the four quarters into which geographers have divided the World. It is probably the most anciently inhabited, and it is to this day by far the most populous in the globe. Throughout the whole extent of this region, from the beginning of time, certain defined characteristics are and have been always apparent. Montesquieu remarks that, while honor is the principle of European monarchies, and virtue the principle which ought to govern republics, whether ancient or modern, fear is the universal principle in despotic Asia. Of the truth of this last statement I think there can be little doubt. Whether in the most eastern portions of China, Siam, or Burmah, or in all the countries under the Mussulman sway, extending to Persia, Turkey, and the shores of the Mediterranean, there is an unvarying uniformity of despotism. Everywhere fear is paramount, everywhere fear is the ruling principle. Asiatic nations never seem to have been inspired by any of the loftier motives which animate Europeans. We cannot recognize among them patriotism, or honor, or moral principle. Their attachment to their religion, whether Buddhist, Hindoo, or Mohammedan, is very tenacious, and would be an ennobling sentiment but that the religions themselves are so inherently false and corrupt. Their treatment of their women appears to be another strong distinguishing trait; everywhere the females are more or less secluded, and have been so from the earliest times. I was reading lately narratives of two English ladies who resided for a considerable period in the capacity of governesses in the harems of two Asiatic Princes at the opposite extremities of that Continent. One lady was engaged to instruct the numerous offspring of the Sultan of Siam, the other the children of the Viceroy of Egypt. Perhaps hypercritics may object that Egypt is not Asia, but Africa. No doubt it is so, but the race belong, by habits, customs, religion, and I believe origin, to the Asiatic family. It is singular to find how closely these two narratives resemble each other.

Another feature of the Asiatic character is its immobility, which pervades all the nations which compose it, whatever their religion. Chinese, Japanese, Persians, Turks, and Hindoos, all adhere pertinaciously to the customs and habits of their ancestors, all resist doggedly the introduction of any change. Their sense of morality is lower than that of the European nations; they are treacherous, dissimulating, and cruel; they evince a great disregard for human life and human suffering; they are not deficient in animal courage, and when trained and disciplined make good soldiers. One of the most remarkable of their characteristics is that they attained, at a very early period, a very considerable proficiency in all the arts of civilization, and have remained stationary ever since. Our recent Exhibitions of Industry in Paris and in London have rendered us familiar with a great variety of Oriental fabrics, and most of us have observed the beauty of the designs and colors, and the excellence of the workmanship of all the manufactures exhibited, particularly of those woven by the hand. Carpets, brocades, and silk stuffs, equal ours in beauty, and surpass them in durability. But the range of these industrial products is very much limited; they are chiefly confined to those produced by manual labor, in which all the Eastern nations appear to possess much delicacy of touch and dexterity of the hand. They have also a considerable talent for imitation, and can copy most of our productions with great fidelity. They seem never to attain to inventive genius, or even in practice or in theory to be gifted with a capacity for Science. They are wonderfully stationary; what they are now they appear to have been, as far as we have any records, two thousand years ago. Arrian's "History of Alexander" describes the natives of India much as they exist at the present day. The History of Asia, too, reproduces at successive epochs the same series of events. The natives of the southern and more fertile and genial regions seem to become relaxed and effeminate; the tribes of Central Asia swell in numbers, and under the command of some great conqueror, Genghis Khan or Tamerlane, fall like an avalanche upon the effeminate natives of the southern regions and extirpate or reduce them to slavery, settle in their country, and are by the operation of the same causes corrupted and rendered effeminate, to be in their turn subjugated by new eruptions from the North.

Asia is certainly the earliest seat of civilization, and, if mankind were all equal in natural powers, would have had the advantage of a start of at least two thousand years. But the Asiatic races, having attained a certain imperfect stage of culture, seem never to have made any further advance. They have been especially unable to realize the fruits which grow from the exercise of the higher mental faculties. It is remarkable that although they have been in constant intercourse with the more advanced European nations, and although they have a natural facility, yet they never have been able to imitate any of our higher productions of Science. For example, English ships have found their way to China ever since the discovery of the Cape of Good Hope, and for the last two hundred years our commercial intercourse with that country has been frequent and increasing. Chinese are active sailors and have long navigated their own waters; it is even said that the Mariner's Compass was known to them before it was introduced into Europe; and yet, though they have our superior ships before their eyes, they have never improved upon the model of their heavy, cumbrous junks. They have never had sufficient enterprise to construct a ship on the European model, or to return to us in Europe the visits we make to them. If Chinese were at all made of the same stuff with English or Americans, we should have had long since Chinese merchantmen visiting the western coasts of the American Continent, from San Francisco to Panama, Callao, and Valparaiso, and perhaps anchoring in the harbor of New York, or in the Thames. Having the European examples constantly before their eyes, they never seem to have made the slightest advance in this or in any other European art. The only exception to this rule of Eastern immobility seems to be found among the Japanese, who of late years have both built and navigated steam-vessels entirely by their own native artificers and engineers. The Japanese are supposed, however, to belong to the Polynesian family.

If we pass in review the long catalogue of inventions, discoveries, and additions to the knowledge and power of Man during the last five hundred years, and which may be said to have exalted his position and to have given him such immense command in the material world, we shall find that not one of them can be traced to an Asiatic origin. They are all the offspring of the higher intellect and genius of the European nations. I know no more convincing proof of the inferiority of the Asiatic mind than this fact. Even with all the example of Europe before them, they have never been able to tread in our footsteps or to follow in the paths which have led us to so high an eminence.

As a general rule, the features of character which I have noticed as pervading the Asiatic races are universal throughout that Continent, and have existed there from the beginning of history. They constitute a type clearly distinguishing them by unmistakable characteristics from the other quarters of the World.

If we turn our eyes to Africa, and examine the Negro race, we shall find them sunk to a much lower level than the Asiatic nations. This vast Continent has been peopled from the earliest times chiefly by the Negro race, and all history represents them as very much in the same low state of civilization which they occupy at present. The Egyptian hieroglyphics represent them in the character of slaves performing the rudest tasks. Travellers have never discovered in any part of Africa buildings, public works, or monuments of any antiquity or durability. In the long series of ages during which they have peopled that Continent, they have left scarcely a trace of their existence. They have no literature, no arts, except of the simplest kind; in size, stature, and muscular strength, they surpass most other races, and their peculiar physical constitution enables them to labor without injury in the hottest regions of the torrid zone. But they are naturally indolent, and do not voluntarily support any toil. Their moral nature appears to be far milder than that of the Asiatic, and though occasionally liable to bursts of ferocity, they seem capable of strong attachment, and not destitute of the gentler affections. Still, whether we contemplate them in their native Continent, or as naturalized in America, they occupy a very low place in the scale of Intelligence.

If we pass in brief review the remaining members of the human family, the savage tribes scattered over different portions of the globe, we shall nowhere discover indications of any great power of improvement. The late Archbishop Whately was of opinion that savage and uncivilized man could nowhere attain to civilization except under the tutelage of a more advanced race, and that generally even with such guidance they were incapable of reaching it.

The history of the North-American Red Indians confirms this view. In a state of nature they are represented as among the noblest of savages, but they have never during three hundred years been reclaimed from their wild lives, or induced to acquire the rudiments even of civilized life. They have never been incorporated in the society of the white man, but have uniformly been driven, as he advanced, farther and farther into the wilderness, and been reduced in numbers, and will probably become shortly extinct. Unlike the Negro they cannot exist in a state of servitude; they must live like the wild animal of the forest or must perish.

Wherever we direct our eyes we are unable to discover either in Asia, Africa, or in any savage life, any equality with the European, or germs of a nature capable of attaining to the same height. It is bootless to try to trace how it is composed or from whence it came. We may have originally been Aryans, or Caucasians, or derived from any other stock, or combination of different stocks, but it is quite clear that we are at this moment the dominant race in the World. Our superiority was visible from the earliest appearance of the race in the records of history. When the Greeks first encountered Darius at Marathon, and Xerxes at Thermopylæ and Salamis, the same superiority of nature was as apparent as when Clive, with his handful of Europeans and Sepoys, defeated the hosts of Surajah Dowlah, and founded our Indian Empire. When Alexander the Great led his victorious bands from the shores of the Mediterranean to the centre of Hindostan the same inherent superiority of race was exhibited.

We recognize the same character of superiority in their Literature. While all Eastern writings are characterized by a certain mystic and obscure mode of expression, the pages of Xenophon and Thucydides are the perfection of style, and speak to us the language of truth. It is marvellous, indeed, to throw a retrospective glance upon the whole body of Grecian literature, and to recall the fact that in an age so remote it constitutes one of the noblest in the World. It attained at once the utmost perfection of style. In the words of Johnson, "the Poems of Homer were only known to pass the common limits of human intelligence, when it was found that age after age and nation after nation could do little more than copy his incidents, new name his characters, and paraphrase his sentiments." It contrasts in this particular with all the nations of the Asiatic Continent. They never appeared to have possessed any body of works worthy to be termed a Literature.

It is not, however, in Poetry alone that Greek writers have attained the highest excellence. Poetry may be supposed to be the first and natural expression to which the efforts of a young nation whose imagination is fresh are directed. But it is not in poetry alone that the Greeks excelled. The works of Plato and Aristotle breathe the profoundest philosophy, and are inspired by deep thought. The Orations of Demosthenes continue to rank as the most finished specimens of eloquence that the World has ever witnessed. In all these departments of literature the Greeks appear to have attained at once perfection. From what source they imbibed their thoughts, or in what school they acquired their inimitable grace and beauty of language, must remain a mystery to us. Of this we may be quite certain—that it was from no Asiatic fountain. The same sudden and immediate attainment of the highest standard of excellence is noticed in Sculpture. If, according to Johnson, succeeding poets have never been able to surpass Homer, the sculptors who have followed his footsteps for 3,000 years have never been able to equal Phidias. Nothing can be more marked than the distinction between the Greek type, and the character of any Eastern nation we have ever known. No progress of civilization, no cultivation of the taste or of the understanding, has ever enabled succeeding nations to surpass these exquisite models.

In Politics, in Laws, and in Government, the European element was equally discernible. The vices inherent in all Democracies soon destroyed indeed these brilliant little Communities; but they exhibited varieties and energies during their brief existence quite foreign to every Oriental nation of which we have any knowledge. The love of liberty or the sentiment of patriotism is utterly unknown to the Asiatic. When Greece was absorbed into the Roman Empire, and that great and glorious nation gradually raised itself to supreme power through ages of sustained effort, the same thoroughly European character was reproduced; indeed, of the two nations the Roman was perhaps the more intensely European. If less brilliant, if less deeply imbued with the sentiment of beauty, his was the more robust and firmer nature. During the long series of ages in which he was gradually arising to supreme power he enjoyed the advantage, which I believe the Englishman alone has ever shared with him, of a political constitution in which the aristocratic and the democratic elements were combined; and it is probably to the durability which resulted from this union that he owes his twelve hundred years of national existence. But all the features of his national character, all the records of his glorious history, bear the unmistakable impress of the European type. We can trace in the Roman qualities akin to our own. We can thoroughly sympathize with his lofty patriotism, and his indomitable constancy of purpose. We feel that he is made of the same materials as ourselves. But we are sensible that we have nothing in common with the Chinese or the Persians at any period of their history. From the earliest origin of the European Nations, at least from the earliest time when any knowledge of their existence has been handed down to us, we find them exhibiting a marked difference from, and evincing an incontestable moral and intellectual superiority over, the older civilizations of Asia. It is not merely that they have surpassed them, but that the whole of their national character is essentially different. This is shown in their writings, in their political institutions, and in their history.

Another great event—the greatest in the life of mankind—served at once to exhibit and to complete the essential difference between them. Christianity had its birth on Asiatic ground, but the seed never germinated there. Any considerable sect of Jewish disciples does not appear to have survived the age of the Apostles. Humanly speaking, we should be led to suppose that had it been confined to the limits of Judea, it would have been crushed by the persecuting influence of the Pharisees. But this precious seed was transplanted to a more genial clime. Blighted in the land of its birth, it sprang up with vigor and luxuriance in the soil of Europe. Asia rejected, Europe claimed it. From the moment that, under the authority of St. Paul, the converts to the new Faith were absolved from obedience to the Levitical Law, the purely Jewish portion were speedily absorbed into the universal character of the new Faith. The difficulties and obstacles which it met with at home were no longer encountered in the new and larger theatre to which it was transferred. The Romans and the Greeks had little intolerance either in their nature or their policy. The fanciful and poetical Mythology which they had moulded into a religion, did not sink deeply enough into their minds to inspire either zeal or enthusiasm, and in the eyes of the philosophic and educated classes was regarded only as a graceful superstition. The readers of Cicero, the disciples of Plato, and above all of Socrates, were not contented with this shadowy and unsubstantial worship. The upper and more educated classes were immersed in the speculations of the Greek philosopher. Far from repulsing with intolerance any form of religion which addressed itself to the understanding, they were quite ready to accept or at least to examine it. The fables of their pagan deities were utterly insufficient to content that craving after religion which is a necessity of our nature. When, therefore, the earnest teachings of St. Paul were addressed to them they met with no unwilling listeners. The temper in which the governor Festus replied to him, "Almost thou persuadest me to be a Christian," is descriptive of the attitude in which the more educated portion of the Greeks and Romans were disposed to meet him. It is, I think, a mistake to suppose that the writings or the influence of the Greek philosophers were hostile to the introduction of Christianity. On the contrary, they prepared the way for it, and there is much in the teachings of Socrates not far removed from the spirit of the Gospel. When, therefore, the new doctrines were preached to them, there were many who were ready to embrace a belief which at once satisfied their understandings and warmed their hearts.

Thus the purifying and ennobling Faith of Christ advanced, until it triumphed in its final adoption by that great Empire which, although verging to decay, was still the Mistress of the World. Its reception by the Northern warriors under whose prowess Rome fell, is, perhaps, more difficult of explanation. In one respect they may have resembled the Romans; they do not appear to have entertained any strong attachment to the gloomy superstitions of their ancestors, or to have numbered among their population any powerful priestly class. Thor and Odin seem not to have been regarded with much deeper respect than were Jupiter and Saturn. They might naturally also have been attracted toward the opinions and faith of a polished people, and might, even while they conquered, still have looked up to them as their guides. Perhaps, also, a more direct, though invisible, influence by the agency of Providence, might have guarded the religion through this period of danger and transition. Whatever may have been the causes, the fact is indisputable, that Christianity, far from sharing the overthrow of the Roman Empire, was embraced by its conquerors, whose power gave it a wider circulation and a new impetus.

While such was the victorious course of Christianity in the Western World, its fortunes in Asia were very different. Although it seems to have become established in the maritime cities of Antioch and Alexandria, which were more Greek than Asiatic in their composition, it never appears to have penetrated far into the interior. It certainly never reached either Persia or India, and whatever footing it may have gained in Syria or Asia Minor was utterly destroyed by the sweeping tide of Mohammedanism. Nor has Christianity, during succeeding centuries, ever accomplished any settlement on the whole of that vast Continent. Buddhist, Bramin, and Mohammedan, have equally adhered with obstinate tenacity to the traditions of their fathers, and utterly rejected all the efforts which the zeal of missionaries has prompted for their conversion. Among all those Religions the only one which assumed an aggressive attitude was the Mohammedan; and it is worthy of remark, that the burst of fanatical zeal, with which the followers of the Prophet were inspired at its commencement, is the only strong impulse which ever appears to have raised the Oriental mind from its passive and inert state. They poured like an avalanche upon Western Asia, wrested all Northern Africa (the land of Tertullian and St. Augustine) from the faith of Christianity, and turned the flank of Europe by their subjugation of Spain. For a moment the fate of the Civilized World seemed doubtful. The Battle of Tours decided the question, and assured the ascendency of Christianity and of Europe. The Moslem tide ebbed as rapidly as it had risen; and it is not without a feeling of surprise that we recall to memory, that, for several hundred years afterward, the Moorish Caliphs retained possession of the fairest Provinces of Spain, and that even to the present day, with the single exception of the recent French conquest of Algiers, the Mussulman continues to rule in the whole extent of Northern Africa, from Morocco to Cairo. The European character appears to have received a new and important element through the bands of hostile warriors who overthrew the Roman Empire. It was the spirit of Chivalry, at the foundation of which lay a high principle of honor, and an intense sense of personal dignity and self-respect. Mr. Burke remarks, that the European nature was composed of two elements, "the spirit of Religion, and the spirit of a gentleman." These seem to have been blended in the composition of these fierce warriors, and to have infused a moral elevation into their otherwise rude and savage character. The Feudal System, assailed by the modern philosophy of the French school, is the source of all the evils which they conceive press upon society. I believe that we owe to it much that is great and generous in our nature, and that to it may be traced much of that spirit of freedom which pervades our modern institutions. "The generous loyalty to rank and sex," in Mr. Burke's immortal passage, embodied two great and fundamental ideas associated with the institutions of chivalry. In a rude age the deference, and almost worship, paid to women, and the customs which placed their weakness under the safeguard of knightly honor, raised at once the tone of sentiment and morals. While all the Eastern nations withdrew their women from all contact with society, secluded them in harems, and sometimes treated them as slaves, the gallant chivalry of Western Europe placed them at the head of society, in the foremost rank of honor, and protected them with the rampart of their swords. Thus from the bosom of these nations, accounted as barbarous by the more polished Greeks and Romans, sprang a purifying and ennobling influence to which they had been strangers. It is in great part through these influences that the germs of a higher civilization were preserved during the long eclipse which preceded the revival of the fifteenth and sixteenth centuries. The great movement of that period was imparted to a society which these influences had rendered worthy to receive and extend it. Four centuries have passed since that great epoch—four centuries which, by their achievements in art, in science, and in learning, have completely changed the position of Man upon this globe. It is needless to enumerate them; they must be present to the minds of all persons of ordinary education. Half the World was then unknown to us: every portion of it has now been explored, surveyed, and mapped. Millions of colonists have planted new empires in what were unknown wildernesses. The Arts and Sciences have explored the secrets of nature, and extended our knowledge to worlds beyond our own. All this vast movement which originated in these centuries is continuing in our own time with unabated force. It seems startling to reflect that, in the latter half of the last century, just a hundred years ago, New Zealand was unknown, and the bare existence, but not the size or extent, of Australia, recognized. It is curious to refer to the charts published previous to the voyages of Captain Cook, and to observe how many blanks on the surface of the globe exist in them at this comparatively recent period, within the recollection of the grandfathers, and even of some of the fathers, of living men. Within the last forty years, as I had occasion to notice in a previous passage, steamboats, and railways, and electric telegraphs, have quite changed the position of man. Our last inventions seem to be of weapons and engines, which have revolutionized the art of war. One effect of these inventions is to place the superiority of Europeans over all the less civilized nations of the globe upon an unassailable height. The suppression of the Indian Mutiny was much to be ascribed to the perfection of our Enfield rifles and improved artillery. The same auxiliaries gave us an easy triumph over the Chinese in the march to Pekin, and the Abyssinians were destroyed by the fire of our troops, without being able to take a single life in our ranks.

Now there is an important deduction to be drawn from this brief retrospect. When we speak of the Progress of Mankind, of the advance of Civilization, we in fact mean the progress of the European portion of the human race, and the advance of Civilization among their populations. All this astonishing advance can be clearly traced to the commencement of the European nations: to the Greeks, to the Romans, and to the European States. No one can suppose that since the imperfect rudiments of civilized society which existed at a remote period among the Asiatic people any substantial progress whatever has been made by any other race, except the European. Of all the great discoveries, of all the inventions which have illustrated this long period of three thousand years, not one can be traced to an Asiatic or still less to any other origin. The Progress of Mankind means the progress of the European branch of the human family.

The most considerable advance in this whole period may certainly be dated from the revival of learning in the fourteenth century; and this is so great that the European nations of the present epoch may almost be ranked as beings of a higher order to those of the preceding periods. This rate of progress seems to be going on in an accelerating ratio, and there is no ground whatever to suppose that it will be arrested. The very increase of power which each new discovery or invention gives to Man is a fresh lever by which he can force his way to another stage of development. There is no ground for supposing that either the resources of Nature or the invention of Man are exhausted. There are probably many secrets within her bosom quite as valuable and available for Man's use as the Magnetic needle or the passage of the Electric fluid along an iron wire. It is quite possible that we may discover or invent a motive power as strong as that of steam, but more portable, cheaper, and more easily adapted to manual use; vapors affording a more brilliant light than gas or lime light may be found out; agents as valuable as chloroform in relieving pain or in curing disease may be discovered. Mechanical inventions are every day produced, abridging labor and extending the empire of Man over the material world. We see that there is still a vast area on the globe unoccupied, or only partly inhabited by Man, and the rapidly increasing population will there find space to expand.

The progress of Man has not been altogether confined to the material world, although I heartily concur in Mr. Buckle's view that abstract Metaphysics have been absolutely stationary, and that we know no more about the nature of the Soul, the principle of Life, or any other question of abstruse Metaphysics than the Greeks did in the time of Plato. There are, however, some branches of moral and political science in which great and important progress has been made. Of these the principal one is the science of Political Economy, which has only existed about a hundred years, as it may be fairly dated to have commenced with Adam Smith. This science has thrown the strongest light upon many questions of politics and of government, and is the right hand of the political statesman. Whether considered in that branch of it which deals with the increase and distribution of wealth, or with that still more important part which relates to the laws regulating population, it affords the clearest insight into the internal mechanism of society. It not only indicates the principles of a sound policy, but it furnishes the correction to a number of those plausible fallacies which are always misleading Mankind. It is unfortunate that while those among its principles which appear to favor popular theories are greedily adopted, the sound Conservative truths of which it is the depository are passed over altogether. The repeal of the Corn Laws was eagerly embraced, and Mr. Bright and Mr. Cobden almost deified; but when the doctrines of Political Economy were applied to demonstrate the fallacies of Communism or the deterioration caused by indiscriminate relief, it met with no such willing converts.

The justice of this view is strongly illustrated at the present moment in the controversies which are raging between Capital and Labor, and in the attempts which are made to alter the natural balance between them by artificial limitations. The leading principle in Political Economy is perhaps contained in the celebrated answer of the French merchants to the Minister Colbert, in the reign of Louis XIV., "*Laissez-nous faire.*" The cardinal article in the creed of the political economist is that all the constituents in the production of wealth find their own level and adjust themselves, and that any attempt to interrupt their natural course is either vain or mischievous. Many of these errors are to be traced to that thoroughly erroneous principle of the natural equality of Mankind, a principle which strongly exemplifies the pernicious effect of a fundamental error in vitiating a whole train of deductions and in rendering a theory thoroughly unsound in all its consequences because of the rottenness which is seated at its core. That we shall find proofs of this in the latest achievements of human progress is as clear and abundant as in the earliest steps of Man.

Let us take that great invention of our own day — the construction of Railways. When the single mind of George Stephenson elaborated the plan of the Manchester and Liverpool Railway, or of the other lines which his prolific and calculating brain perfected, he worked alone. The conception was the produce of his powerful intellect; and if to him we add the names of Brunel and two or three others of the greatest and most eminent engineers, we shall find that the whole Railway system of England was put in motion by a very few master-minds. Yet what a prodigious amount of labor was required to execute them! How many navvies have toiled in tunnels or in cuttings and embankments! We have most of us probably watched the progress of some of these works; there is a set of powerful men in their shirts with their bare muscular arms engaged in filling wheelbarrows with soil, trundling them along a narrow plank, tilting them over at the end of it, and returning with their barrows to repeat the same operation. It is difficult to conceive any work of man requiring less the use of his mental powers. The operation is merely mechanical, and I am not sure if a well-trained orang-outang might not be competent to perform it. Nothing is required but bodily strength; a navvy need not write or read or know his multiplication table; education, as far as the performance of his task is concerned, is useless and superfluous. In the construction of one of the great lines of railway thousands of men are so employed for months and even years. Without their aid the Railway could never have been made, and the mental creations of Stephenson or Brunel would have remained as unsubstantial as the baseless fabric of a vision, and yet all this amount of manual labor has been set in action by the mind of one man. I think we may venture to extend the application of this example, and to lay it down as a principle that the progress of Science in its application to human wants augments the demand and necessity for manual labor in a far more rapid ratio than it does the demand for intellectual qualifications. The ingenuity of Walton and Arkwright, and the high mental power engaged in the construction of their machines, abridged the cost of production and gave room for the employment of millions of hands in the manufacturing industry. Yet what can be more thoroughly unintellectual than the occupation of a weaver sitting at his loom and passing years in sending his shuttle to and fro? Those modern disciples of progress who conceive that they have in Education a panacea for raising all Mankind to the same level, forget that acquirements when not used are lost. A weaver can move his shuttle, a navvy can fill and empty his barrow, without the slightest occasion for, or in deed the possibility of using the arts of reading and writing. What is never used is speedily forgotten, and although good fortune or superior natural ability may enable an individual here and there to profit by those acquirements, yet the probability is that a very small residuum will remain to the masses after a few years of toil in the performance of these mere manual tasks. Yet it is imperative that they should be performed or the civilized world would perish, and the truth becomes apparent, that the amount of mere manual labor, requiring the smallest amount of mental capacity, increases in a far more rapid ratio than do those offices which demand even a very moderate exercise of the intellectual faculties. The tendency of the advance of Civilization is rather to widen than to lessen the lines of demarcation, which divide the possessors of intelligence and capital from the mere laboring class, and all attempts to efface these distinctions by the operation of compulsory laws and artificial restrictions may retard the march of Progress, but will inevitably fail of their object. The political economist is no friend to a crowded statute book, he has little belief in the efficacy of compulsory enactments or arbitrary interferences with the natural course of events. He appeals to the experience of the last four hundred years, and points out that while Governments have been occupying themselves with little effect, or with an injurious effect, in legislating on measures which have chiefly for their object the possession or the transfer of political power, those great social movements, that mighty spread of scientific knowledge combined with its practical application, which have changed the face of the world and the position of Man on it, have gone on quietly in effecting their wonderful changes unperceived or disregarded by the rulers of nations.

The historian who should hereafter narrate the events of the present century would, if he followed the example of his predecessors, dwell with minute accuracy upon the historical events which have been crowded into this remarkable period. He would describe the whirlwind of the first French Revolution, he would delight to dwell upon the astonishing career of the First Napoleon, he would narrate the successive Revolutions which have convulsed France, and would analyze the provisions and endeavor to trace the effect of our own two Reform Bills, but he would probably pass over with a cursory notice the invention and application of Steam, the Electric Telegraph, and all the other marvels of Science, which have transformed the whole state and condition of Man, and effected a change as complete as if a new living being had been created. These changes steal upon us im perceptibly, they are heralded by no fierce contests for power, they involve no sudden destruction of whole classes, no violent convulsions of society; at their commencement they scarcely attract notice, they steal on gradually, and it is not till their operation has become felt in its extended effects that the immense influence it is exerting upon the destinies of Mankind become slowly apparent. There is another happy characteristic in this march of scientific progress—that in its results it is always beneficial. It is perpetually adding to the aggregate of human power, wealth, and enjoyment, and it never reaches its end through human suffering. It uses no guillotine, it establishes no conscription, it plunges into no wars, it prosecutes its peaceful contests by the universal assent of all populations. We are not aware of the benefits it is confer ring upon us till we suddenly awake and find ourselves richer, wiser, happier and better than we were.

The discoveries and the progress of the last four hundred years have raised the human race (or at least the European branch of it) to an elevation so infinitely superior to that of any other known living being that it is not presumption to infer that he is the main object of God's providence on earth. Nowhere do we see any rival to him, and the whole scheme of creation, as far as this planet is concerned, appears to have him for its end and object.

There is a tendency which is much to be deprecated in modern Religion and in modern Science to diverge somewhat from each other. Religion perhaps inclines to become narrow, exclusive, dogmatic and polemical; while in Science there is a disposition to refer every thing to secondary causes and to reduce Religion into a cold abstraction in the place of a warm and living sentiment. Each of these tendencies is to be deplored.

The discoveries of modern Science by their comprehensiveness afford a Key to enable us better to conceive the mighty scheme of Providence; the coherence of the different parts, the harmony of the whole, the adaptation of means to an end—afford additional proofs of the unity of the design and of the Wisdom, Power, and Goodness of its Great Author. Let us suppose one of the Sages of antiquity, or one of the early Fathers—Socrates or St. Chrysostom—tracing out or trying to draw the plans of Omnipotence. How scanty and imperfect would have been their materials, how unconscious they would have been of all the manifestations of the Deity displayed in the Solar System; how ignorant of the wonders of chemical and mechanical Science, and of their applicability to the wants of Man; how ignorant of the Geography and of the various conditions of this globe itself; how totally unacquainted with all the secrets of antecedent states of the Earth which Geology dimly and imperfectly reveals to us—how difficult for them to have proved the vast capabilities of Man, or his steady progress to a higher grade in the scale of the Creation, which his acquirements during between two and three thousand years have proved. Even Paley, writing so near our own times that he is almost a contemporary, wanted the proofs in favor of his principles of Natural Theology which these later times would have afforded. It is as if some great and glorious landscape (we will say the first view of Switzerland and the Alps from the descent of the

Jura above Jex) was gradually revealed to the traveller by the dispersion of the morning mists. First, he would catch the rich and varied landscape at his feet, then gradually the bright, crystal reflection of the Lake of Geneva; slowly and imperceptibly would he discern the grand and picturesque forms of the lower ranges of the mountains of Savoy; and, lastly, scarce believing his eyes, scarcely crediting that the giant glittering masses piercing the skies belonged really to this earth, he would gaze upon the sublime spectacle of Mont Blanc and the highest range of the Alps. Thus we of the nineteenth century have unveiled before us much of that mighty scheme which earlier generations could never have suspected. We are enabled in some degree to measure its grandeur, its immensity, and, above all, its Unity. We must feel convinced that to Astronomy, Geology, and the whole circle of the Sciences—from those which can by Microscopic power scrutinize the minutest object in terrestrial economy to those laws which embrace the Universe—the same stamp of Omnipotence is affixed. We may gain additional ground for the belief that the Human Mind, which has been so cherished and favored by its Divine Creator, may find addi tional confirmation of those hopes which He has given us, that our living participation in this great scheme does not end here, but that we shall be sharers in the destinies of that Immortal Creation of which it is given to us at this advanced stage of our progress to know so large a portion.

THE END.

www.ingramcontent.com/pod-product-compliance
Lightning Source LLC
LaVergne TN
LVHW011234110826
845150LV00006B/1632

9781425514815